Tuning for Economy

Tuning for Economy

COLIN CAMPBELL
M.Sc., C. Eng., M.I. Mech. Engrs

LONDON   NEW YORK

Chapman and Hall

First published 1981
by Chapman and Hall Ltd
11 New Fetter Lane, London EC4P 4EE

Published in the USA
by Chapman and Hall
in association with Methuen, Inc.
733 Third Avenue, New York, NY10017

© 1981 C. Campbell

Typeset by Inforum Ltd, Portsmouth
Printed in Great Britain by
Richard Clay (The Chaucer Press) Ltd., Bungay

ISBN 0 412 23480 7 (cased)
ISBN 0 412 23490 4 (paperback)

British Library Cataloguing in Publication Data

Campbell, Colin
  Tuning for economy
  1. Automobiles – Motor
  I. Title
  629.2'504   TL209

  ISBN 0-412-23480-7
  ISBN 0-412-23490-4   Pbk

Contents

Preface

This book is for the typical motorist who, as shown in the first chapter, only looks at his engine when he remembers to check the oil level or has noticed a tendency to misfire under load. He could be aware that the car no longer gives the good mpg it gave when he bought it as an 'immaculate used car' and could be wondering what to do about it. This book will tell him. The richer motorists in their new Jaguars and Mercedes – the ones that flash past us on the Motorway – never worry about fuel consumption. This book is not for them.

Here, then, is an explanation of the principles of tuning, followed by a simple step-by-step DIY tuning schedule. A hard look is taken at the many magical economy devices offered in the Press. Two of these are shown to stand up to critical examination and honest testing, namely some electronic ignition kits and the Kenlowe electric fan. Two more devices show marginal gains in economy. Finally, we offer advice on how to approach the economy driving techniques of the successful Economy Rally drivers.

Since we live in such a cynical age we find it necessary to state that we have no connections with, nor have received any financial inducements from, any of the recommended firms making tuning equipment or economy aids. We must thank these companies for all the useful information they have supplied. A full list is given in the Appendix.

We would like to thank Champion Sparking Plugs Ltd and the Automobile Association for their valuable advice and assistance.

Suffolk   C.C.
1981

1

The Fuel We Waste

Even though the word 'inflation' will raise a yawn in any company today we cannot escape the consequences of the alarming price increases in World oil supplies in recent years. Even an index-linked salary would not protect us since oil prices have inflated at a greater rate than other commodities. In March 1970, 4 star petrol in Great Britain cost 33 pence per gallon. By March 1980, the price at the pump had risen to £1.30. A similar rate of increase in this decade would give us the £5 gallon by 1990.

Our state of tune

By 1978, The Champion Spark Plug Company had completed a comprehensive survey on the state of tune of more than 13 000 petrol-engined vehicles in the U.S.A., Canada, Mexico, Great Britain and four other European countries. The tests showed that 84% of all vehicles tested had at least one defect contributing to reduced fuel economy. Of the European cars tested, 54.5% had incorrect ignition timing and 76.5% needed carburetter adjustment.

The test equipment was as comprehensive as that used by the most modern professional tuning establishments. There was an initial examination bay, a diagnostic bay and a rolling-road dynamometer to measure the brake horse power transmitted to the road surface by the driven wheels. Fuel consumption was measured on the dynamometer before and after receiving a full tune-up. The percentage improvement after tune-up was fairly consistent for three European countries; 12.8% in Belgium, 11.7% in Great Britain and 12.2% in West Germany. In Spain it was 9.2% and in Italy only 4.6%. The latter figure is no doubt proof of the Italian male's passionate love affair with his automobile. They do spend a lot of time with their heads buried under the bonnet.

The evidence suggests we have a tendency to neglect our engines in Great Britain. Of the London cars, 30% were found to have the ignition over-advanced, 13% had the vacuum advance inoperative and 12% had the mechanical advance inoperative. One should remember that these two automatic advance systems were developed primarily to improve the fuel economy of the automotive engine!

Even the more modern cars showed signs of neglect. 10% required new plugs and 24% had badly worn distributor points.

The wasted money

We should note that 84% of all cars tested had defects leading to reduced mpg. 16% were therefore well tuned. Applying a correction for this well-tuned minority means that the badly tuned cars in Great Britain showed an improvement of $11.7 \times 100/84 = 13.9\%$ after tuning. Taking this as representative of the badly tuned cars of this country, the potential saving from regular tuning alone on a car that uses 500 gallons per annum would be £90. Even the cost of professional tuning at regular intervals becomes economically viable. The aim of this book however is to show you how to tune the engine without any professional help. Moreover, by an initial outlay of about £100 on tuning equipment and bolt-on tuning aids we intend to demonstrate how you can save as much as 25% of your total fuel bill!

All this is saved by adopting the following procedure.

(1) Re-tune the engine every 8000 miles.
(2) Fit one of the recommended electronic ignition systems.
(3) Fit a thermostatic electric fan.
(4) Always re-fill the petrol tank early in the day.
(5) Always drive by the recommended economical technique.

1.1 Tuning principles

In an earlier work, *The Sports Car Engine,* the author gave some thought to the price we pay by running a car in a poor state of tune. Using more recent figures, it will be enlightening to consider the overall running costs of two cars, otherwise identical, but with the first in excellent tune and the second in a poor state of tune. This second car will still start without difficulty, will run without misfiring and the driver will not be aware that the power has gradually been reduced since the car was last tuned. Let us suppose that this second car has a dirty air cleaner, an ignition timing that has drifted about 5 degrees from the correct setting and a vacuum advance pipe that has become disconnected from the distributor.

An increase in pressure drop across the air cleaner results in a reduction in density of the air entering the carburetter. This produces an automatic enrichment, the extent of which can increase fuel consumption by 10% *at a given throttle opening*. Since the choked air cleaner offers resistance to air flow a larger throttle opening will be required on this second car to produce the same power as on the first car with a clean air cleaner. The incorrect static ignition timing will reduce the power *at a given throttle opening* by 7–8% and the inoperative vacuum advance mechanism will ruin economy under cruise conditions. The overall increase in fuel consumption will be about 25%.

1.2 The economics of tuning

In Chapter 5 we intend to lead you step by step through a simple DIY tuning technique. Even if you decide this is not for you, it is still economic to keep your car in a good state of tune. Perhaps you are too busy house-painting, double-digging the garden or even enjoying yourself. Perhaps you admit to yourself a natural tendency to fall down on most DIY jobs. No matter! You will still save money even if you have to pay to have your car tuned.

Let us assume that drivers of Car A and Car B each cover an annual mileage of 15 000 miles. In perfect tune, their cars would average 30 mpg. At £1.30 per gallon this would cost £650 for fuel per annum. The driver of Car A has his car tuned by a specialist tuning garage every 7000 to 8000 miles at a cost of £15 plus £5 for tuning spares (plugs, contacts, etc.). This would maintain the engine in an average state of tune to give 97% of the mpg given immediately after tuning, i.e. a falling off of 6% before the next tune-up. The driver of Car B is less fussy and at the end of 15 000 miles, when he notices the engine is running badly, he pays £15 for a tune-up, plus £5 for tuning spares. At the end of 15 000 miles, however, the petrol consumption has increased by 20%, an average increase of 10%.

The driver of Car A will pay £670 for petrol plus £40 for tuning, a total of £710. The driver of Car B will pay £715 for petrol, plus £20 for tuning, a total of £735. The first driver has not only gained in reliability, improved power and acceleration when necessary, but he has saved about £25 in running costs. Finally, the driver of Car C, who tunes his own car only spends £10 per annum for tuning spares and therefore spends £55 per annum less than the driver of Car B. Need we say more?

1.2.1 **The conversion of fuel into useful work**

If we use the word 'thermodynamics' the layman sometimes takes fright. The mathematicians like Carnot and Rankine who revelled in the science when it was new only helped to make it a very esoteric subject. They liked the concept of a 'perfect working fluid' (and a petrol–air mixture is far from perfect) and they suggested, rather hopefully, that we would need to extract every scrap of thermal energy from this working fluid, right down to a thermal state of *Absolute Zero Temperature* to achieve a Thermal Efficiency of 100%. Popular journalists sometimes get hold of thermal efficiencies expressed in this way and make naive statements in the Press and on TV that a modern car 'wastes' about 75% of the energy in the fuel. Even the most modern oil-fired power stations they tell us waste about 65% of the fuel. It all sounds very impressive in a Tabloid newspaper but the concept is a long way from reality in this century. An engine that exhausted its working fluid at a temperature that would freeze liquid nitrogen would obviously create problems in a crowded city street!

Let us try to approach things from a practical angle. The fuel technologist measures the available energy of a liquid fuel by burning a known mass of fuel in a bomb calorimeter. For a typical petrol this value would be about 19 000 BTU/lb (44 200 KJ/kg). Since the experiment is carried out under controlled conditions this value represents the amount of energy we would extract from

burning the fuel in an engine in which no heat is wasted in cooling the combustion chambers and cylinders, no friction losses
occurred in the mechanical components, no work is expended in
inducing the fresh charge into the cylinders and no work needed
to exhaust the burned products. Finally, we are asked to accept
that the final temperature of the exhaust products is room temperature, i.e. 20°C.

With this concept of a 100% efficient engine we would expect to
achieve an overall fuel consumption of about 120 mpg on a
typical medium-size saloon. Unfortunately, we have not yet overcome the practical difficulties of burning the fuel in an insulated
combustion chamber or any of the other assumptions stated
above and a typical modern engine only achieves a mean efficiency of about 25% over the whole range of its operating conditions. The Diesel engine is known to be outstandingly efficient,
approaching an efficiency of 40% at times, and it is possible we
shall see more of these in use in the future. Our concern, in this
book, is the common petrol engine, the engine we are stuck with
since we cannot afford to trade in our current vehicle for a much
more expensive Diesel model.

Fig. 1.1. illustrates the losses we would expect to find in a
modern petrol engine when cruising at 60 mph. At lower speeds,
combustion would be less efficient and the thermal efficiency
would fall.

1.3 The four phases of tuning

Our approach to tuning is given by the cycle of operation we all
know as the four-stroke cycle. Two-stroke engines in cars are so
rare as to warrant their neglect in this book. Logically, then, we
can think of the four strokes of the operating cycle as phases in
which we can influence the overall efficiency of the engine by
careful tuning. The four phases are therefore Induction (including Carburation), Compression, Combustion and Exhaust.

1.3.1 Induction

Two basic requirements exist for good induction. Unfortunately,
they are not compatible and a compromise is necessary on most of

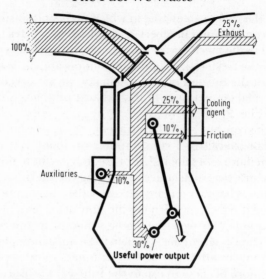

Fig. 1.1 Where the energy in the fuel goes.

the design parameters. The designer of a racing engine sacrifices economy to obtain the highest possible power. Good breathing, or volumetric efficiency, as the automotive engineer calls it, is one of our requirements. The second is good metering, or carburation. In our case, we must strive to achieve the latter at all times without sacrificing too much power with a carburation system that produces a serious drop in volumetric efficiency. When an engine is running at the usual cruising speed the time available when the inlet valve is open in any one cylinder is very, very short. At the modest speed of 2000 rpm the whole induction process must be completed in about one-hundredth of a second. Inevitably, work is expended as the piston moves down and creates a depression between the cylinder and the three main restrictions, the passage through the valve port and the two restrictions at the carburetter, i.e. the venturi and the throttle-plate. With a system that breathes well this depression is low and the power loss is at a minimum.

The second requirement is the precise metering of the *correct* mixture strength to every cylinder. 'Correct' is stressed since we cannot use a weak (or lean) mixture all the time. A richer mixture is required at full throttle since an oxidizing weak mixture would endanger the exhaust valves and an even richer mixture is required when starting from cold. Metering is a large subject and involves the working principles of several proprietary makes of carburetter. A few cars today are fitted with petrol injection systems such as the Lucas system on the Jaguars, controlled by an electronic 'magic box' and the Bosch 'K-Jetronic' system on the Mercedes and Porsches. These systems are far too sophisticated for the DIY tuner. They either fail completely – fortunately a very rare occurrence – or require professional tuning every 20 000 miles. If you own a modern Mercedes you should be able to afford this.

1.3.2 Compression

The compression ratio is fixed and is chosen by the manufacturer for the particular model. It can be increased by machining a slice off the cylinder head gasket face or by fitting special pistons, but these are expensive tuning modifications and do not come with the remit of this book. The compression ratio is a volumetric ratio, not a ratio of pressures. It is the ratio of the free volume inside the cylinder at bottom dead centre (BDC) to that at top dead centre (TDC). By free volume we mean the actual geometric volume enclosed by the boundaries of the cylinder walls, the cylinder head and the piston crown. The actual compression pressures reached when the engine is running will be influenced to a large extent by the volumetric efficiency of the particular design. For example, a high performance sports car such as a 12-cylinder Ferrari fed by a battery of Weber carburetters will often have a quite modest compression ratio. The 400GT model, for example, with 6 Weber carburetters has a compression ratio of 8.8:1. The Allegro 1500, with a push-rod ohv 'cooking engine', fed by a single SU carburetter has a compression ratio of 9.0:1. Such an engine, with its relatively small inlet valves and single carburetter, will only produce compression pressures about 70% as high as those in a Ferrari at the same engine speeds.

Considering the state of tune of an engine there are only three ways by which it can lose compression, omitting the rare case of a loose sparking plug. First, the gas can be escaping past badly seated inlet or exhaust valves. Secondly, it can be leaking past the piston rings. Thirdly, there can be an escape of gas at the cylinder head gasket, a condition which deteriorates rapidly into a burned gasket. There are ways of detecting loss of compression available to the home-tuner and techniques to identify the source. These techniques will be described in Chapter 4. At this stage, it is enough to remember that an engine with one or more cylinders having a poor compression can never be tuned to be economical. Mechanical excellence is all part of good tuning.

1.3.3 Combustion

This is the third stroke in the cycle, perhaps the most critical and certainly the subject of more research and more technical papers than the other three strokes of the engine. In theory, we would achieve the maximum combustion efficiency by burning all the fuel instantaneously at TDC. In practice, there would be an almighty explosion that would blow off the cylinder head. To avoid this disaster and the related problem of knock, the combustion chamber in a modern engine is designed to carefully regulate the rate of combustion and, by virtue of this, to control the rate of pressure rise. This is no easy matter. To delay combustion of a large part of the mixture until the piston has descended to half-stroke would give very inefficient combustion. In a typical engine running efficiently, combustion of 75% of the fuel is completed before the piston has descended 20 degrees past TDC. The remainder is burned by about 35 degrees after TDC. Within the limits imposed by the anti-knock properties of the fuel and the permissible rate of pressure rise this represents the closest approach we can hope for of reaching the combustion efficiency of instantaneous combustion at TDC. We owe this to two generations of engine research. As the more or less grateful recipients of all this work to give us a very efficient engine all we can do at our level is to maintain the pistons, rings and valves in good condition and to tune the ignition system to give ignition at the right time and with unfailing regularity. This in itself is no mean achievement as we shall learn in the next chapter.

1.3.4 **Exhaust**

So long as the exhaust valves open at the right time and the exhaust system does not offer excessive back-pressure against the discharge of the gases, it is inevitable that the majority of the burned products of combustion will escape to the atmosphere since they are at a relatively high pressure. Many years ago, it was realized that the discharge of these gases can be used to drive a gas turbine, this turbine being used to supercharge the induced mixture and thus obtain more power. It has also been known for some time that the energy contained in this hot pressurized gas can be used to assist the flow of the next cycle induction process by means of an extractor action. Under full load the pressure in the cylinder when the exhaust valve opens will be in the region of 60–90 lb/in^2 (4.2–6.3 kg/cm^2). The bulk of this gas passes the exhaust valve at the speed of sound [about 1700 ft/s (520 m/s) at the prevailing temperatures and pressures]. It helps to think of the exhaust gases leaving the cylinder like a fireball shot from a gun. This fireball tends to leave a depression behind it. With suitable valve overlap (the number of crank angle degrees between the opening of the inlet valve and the closing of the exhaust) a good engine designer can use this exhaust depression to help the start of induction and thus improve the volumetric efficiency. The amateur engine tuner cannot easily experiment with valve timings, but he can sometimes improve economy at cruising speeds by a change in the total exhaust system length.

As a matter of routine tuning there is little we can do about the exhaust phase. We can check that the exhaust valves are seating well, are opening and closing at the recommended timings and that the clearances are correct. If the exhaust system is in good condition, with no broken baffles inside the silencer, the tuning programme is completed.

2

Ignition

The well-established normal coil ignition system consists of a battery (charged by an alternator or d.c. generator), an ignition switch, an induction coil, a contact breaker (contained in the distributor body), a distributor and a set of spark plugs. The elementary circuit is shown in Fig. 2.1. The induction coil is the heart of the system. From elementary physics, we know that an electric current can be produced when a loop of wire is moved through a magnetic field. In an ignition coil the opposite occurs; the loops of wire, thousands of them, remain stationary, while the magnetic field changes rapidly. The coil consists of a laminated soft-iron core with two separate concentric windings. The primary winding consists of about 200 turns of relatively thick wire. The secondary winding is of about 20 000 turns of very fine wire. The primary and secondary windings are of course well insulated against internal short-circuits (see Fig. 2.2).

Current flows through the primary windings as soon as the contacts close. This current creates a strong magnetic field around the two concentric windings. When the contacts open the magnetic field collapses and produces an induced current in the secondary winding. The current is minute, a small fraction of an amp, but the voltage induced depends largely upon the rate of change of magnetic flux. By design, it reaches about 25 000 V.

This collapse of the magnetic field also induces a transient current in the primary windings. The e.m.f. in this winding rises to about 250 V and, with no provision to 'store' it, this voltage would be high enough to cause a spark to jump across the contacts. This electrical charge is stored or contained on the surfaces

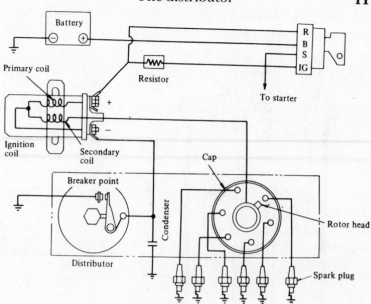

Fig. 2.1   The conventional ignition system.

of a condenser. The condenser is given a high capacity in a small volume by winding a length of metal foil inside a cylindrical container, the foil being coated with an insulating material adequate to prevent any short-circuiting. Not only does the condenser prevent sparking across the contacts, which would produce rapid erosion, but it ensures complete collapse of the magnetic field around the windings by inducing a small field in the reverse direction.

2.1 **The distributor**

The purpose of the distributor is to distribute the high voltage secondary current to the plugs and to ensure that they receive these induced current pulses in the right firing order and at the right timing relative to TDC. Early in the history of coil ignition development, the contact breaker and the distributor were combined into a single unit. This was a natural development since

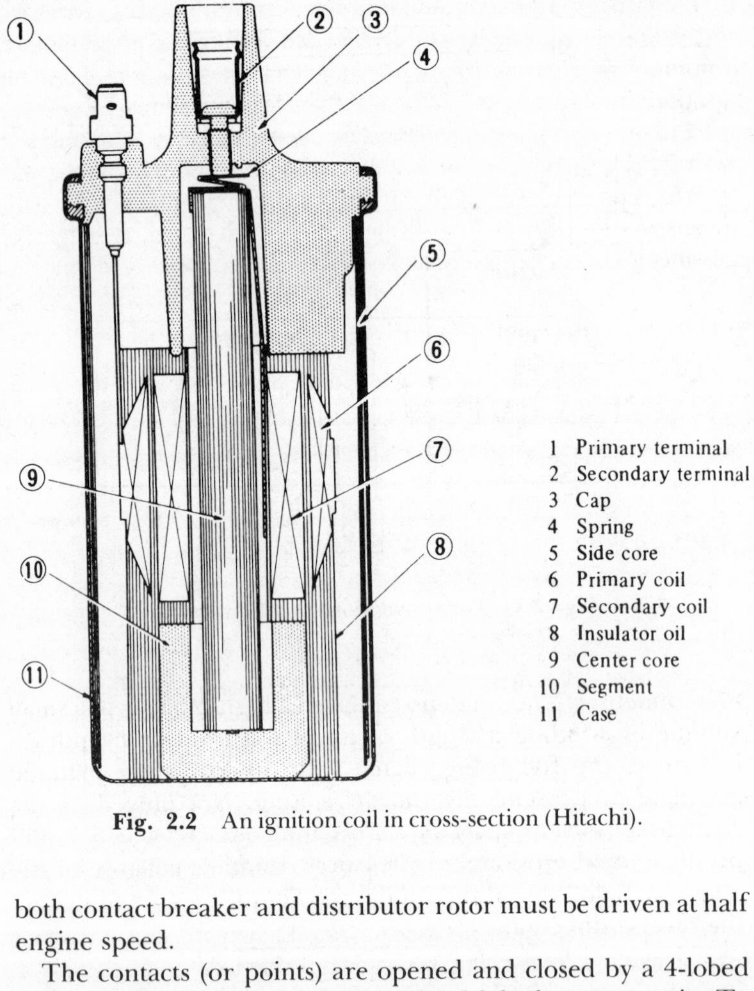

1 Primary terminal
2 Secondary terminal
3 Cap
4 Spring
5 Side core
6 Primary coil
7 Secondary coil
8 Insulator oil
9 Center core
10 Segment
11 Case

Fig. 2.2   An ignition coil in cross-section (Hitachi).

both contact breaker and distributor rotor must be driven at half
engine speed.

The contacts (or points) are opened and closed by a 4-lobed
cam on a four-cylinder engine and a 6-lobed cam on a six. To
achieve efficient combustion under different running conditions
it is necessary to be able to vary the timing of the ignition. For
example, at cruising speeds with a moderate throttle opening the
spark should occur relatively early for maximum economy; when
climbing a steep hill at a lower engine speed with a fully open
throttle the ignition timing must be much more retarded to avoid
knock. Every operating condition experienced by every partic-

ular engine design has an optimum ignition timing for that
operating condition. It is now possible on the more expensive
automobiles, such as the latest 7 Series BMW, to regulate the
ignition timing by means of an electronic brain. Until we can find
the £12–15 000 for a car with such sophistication, we must survive
with the cheaper ignition system that gives a fairly satisfactory
approximation to the ideal ignition curve (or family of curves) by
means of two devices, the centrifugal advance and the vacuum
advance.

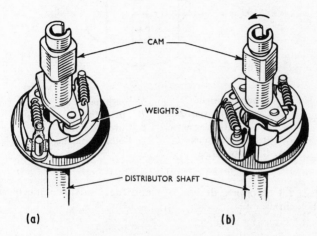

Fig. 2.3   Early design of link-type centrifugal advance mechanism.
(a) Static position. (b) Fully advanced position.

The centrifugal advance, as shown in Fig. 2.3, uses the action of
centrifugal force acting on two spring-loaded governor weights
to rotate the cam plate. A line drawing of a more modern design is
shown in Fig. 2.4. In this Hitachi distributor the governor
weights, 1, drive the cam plate by means of pegs working in radial
slots. Springs, 4 and 8, control the outward movement of the
weights. The amount of centrifugal advance is determined on the
dynamometer by the manufacturer.

In a similar way, the contact breaker baseplate can be rotated by
a side arm connected to a vacuum control unit (see Fig. 2.5). The
flexible diaphragm inside this unit is moved against a control

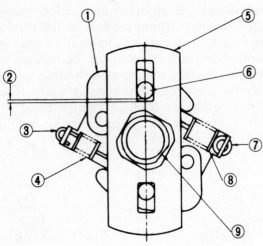

1 Governor weight	6 Weight pin
2 Clearance for start and	7 Circular hook
end of advancing angle	8 Governor spring
3 Rectangular hook	9 Rotor positioning tip
4 Governor spring	
5 Cam plate	

Fig. 2.4 Modern design of centrifugal advance mechanism (Hitachi).

spring by the pressure in the induction manifold. On the majority of engines, the small bore pipe connecting the diaphragm chamber to the induction manifold leads to a small drilled passage in the carburetter body that is covered by the throttle-plate edge when the throttle is in the closed (idling) position. Vacuum advance is therefore inoperative at idling speed. As the throttle is opened the drilled passage is uncovered and the manifold depression is transmitted via the connecting pipe to the vacuum unit. The overall effect is therefore:

(a) Fully retarded ignition, i.e. the static timing value, at idling.
(b) Full vacuum advance by as much as 10° before the static timing when running light at 800 to approximately 1200 rpm.

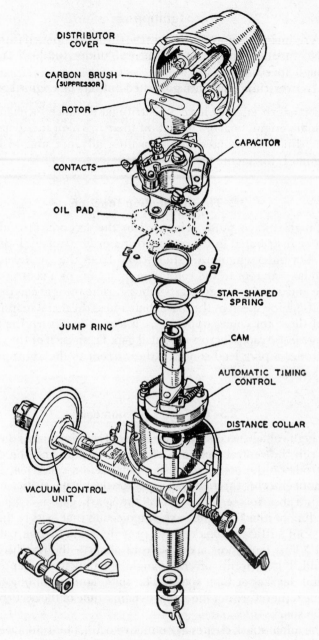

Fig. 2.5 Exploded view of Lucas distributor.

(c) A reducing vacuum advance as the throttle is opened further.
(d) No vacuum advance at full-throttle under full load and at high speed.
(e) An over-riding advance governed entirely by engine speed.

All modern engines have a centrifugal advance mechanism, but some engine makers, based on their work on the dynamometer, find the provision of a vacuum advance unit to be a marginal refinement.

2.2 The distribution function

The high-voltage pulse generated in the secondary windings must be distributed in the correct firing order (usually 1–3–4–2 in a 4-cylinder engine). A short cable from the coil takes the secondary current to the rotor arm by means of a central electrode and a spring-loaded carbon brush. The rotor arm is driven at half-engine speed by the drive shaft and as the metal strip at the tip of the rotor comes opposite each electrode in turn the high voltage pulse jumps across the small gap. From each of the outer electrodes a plug lead conducts the current to the appropriate spark plug.

2.3 The spark phenomena

The high voltage necessary before current can flow across the gap between the central electrode of the spark plug and the earth electrode on the body involves a phenomenon called *ionization*. Without entering this field of physics too deeply it is sufficient for our purpose to see ionization of the spark gap as a kind of breeding of ions from the oxygen molecules that exist in this air gap until sufficient ions exist to provide a conducting path to earth. This ionization takes place in about a thousandth of a second, but in terms of crank-angle rotation it can represent several degrees at high speed. The static ignition timing of an engine is therefore not the actual dynamic time of the occurrence of the spark.

The initial spark creates a conductive path. This is followed by an arc. Many of us are familiar with the arc created between a welding electrode and the metal to be welded. In the case of a spark plug discharge, the arc is of very short duration. The first phase is called the *capacity component* and lasts no longer than 100 μs. The second phase, called the *inductive component,* lasts about twenty times as long and contains the bulk of the electrical energy. Even so, the rate of energy discharge is higher during the first phase, and this is an important factor. The *rise time,* i.e. the time to reach maximum voltage is a critical factor when attempting to produce a good spark across a fouled or dirty plug. A badly fouled plug has a leakage path to earth down the insulator of the central electrode. When the rise time is too long this leakage path can prevent a successful build-up of voltage across the plug points. Let us remember though that we should not be running with dirty plugs!

It can be shown experimentally that the first component produces the spark with a normal plug and this spark is sufficient to ignite the petrol/air mixture adjacent to the plug points. The inductive component produces more light and heat, but with normal conditions existing in the combustion chamber the inductive component contributes nothing to the initiation or spread of combustion. There are occasions, however, when the heat energy of the inductive component becomes important. A weak mixture is slow to ignite and the longer duration of the inductive component is sometimes required to give reliable ignition. A wet mixture, as when starting from cold, requires much more heat pumped into it to vaporize the mixture between the points and then to raise it to the ignition temperature. In such conditions, a large inductive component succeeds in igniting the mixture and an ignition system that produces a relatively small inductive component would fail under such conditions. When we see the thread of flame jumping across the plug gap, this visual witness of the electrical discharge is produced by the inductive component. Thus, the insistence on a *fat* spark by the older breed of mechanic is really an insistence on a large inductive component.

2.4 **The spark plug**

Can the reader imagine that he is trying to persuade a hard-nosed Yorkshire businessman that he can sell him an electrical spark ignition system to replace the older hot-bulb system that will ignite the petrol/air mixture at all times in a combustion chamber with dirty oil-splattered walls and will even keep operating in winter with ice-cold metal surfaces and with a petrol/air mixture that is partially composed of wet droplets? Need we elaborate on the scorn and sarcasm this crack-brained idea would provoke!

If it needs as much as 12 000 volts to produce an effective spark how can you provide a good enough insulator in such damp, and dirty conditions? In any case, everything in the combustion chamber gets covered with carbon and any idiot knows that carbon is a good conductor of electricity!

Fortunately, we have had about three-quarters of a century to concentrate on these problems and we can now say, modestly, that it works most of the time. We try to avoid the worst conditions by improvements in engine design. Nothing is perfect, but the modern spark plug, backed up by a modern coil ignition system is reliable enough to satisfy a Bradford woollens manufacturer. Not many of them even think to carry a spare set of plugs in the boot.

Essentially, a spark plug is a steel body incorporating an earthed point and a central insulated electrode to which is attached the plug lead from the distributor cap. A typical plug is shown in cross-section in Fig. 2.6. The chief problem that challenges the plug designer is to maintain the temperature of the central electrode insulator within the working range. If the temperature is too low a deposit of conducting carbon soon builds up over the surface of the insulator and – as our Yorkshire businessman would soon tell us – the current will flow down the insulator to earth and fail to jump the gap. If the temperature is too high the electrode itself becomes overheated and becomes a hot-spot to induce pre-ignition, i.e. ignition too early. This can be dangerous, giving excessive combustion pressures, overheating, and eventually engine failure. The 'self-cleaning' temperature range for a typical plug is approximately 450° to 650°C.

The plug designer uses three dimensions to provide a full range of plugs to suit all types of engine, from low compression engines burning low octane petrol to high compression racing engines. These dimensions are illustrated in Fig. 2.7. The heat picked up by the central electrode and its insulator during combustion is largely dissipated to the cylinder head, and thence to

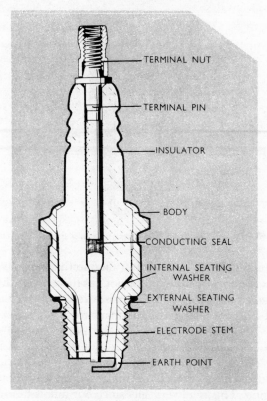

Fig. 2.6 Cross-section of a spark plug.

the cooling water, via the washer that serves as a gas seal for the insulator. It is obvious we cannot hope to use the same plug in a Mini with a maximum power output of 40 bhp per litre and a Lotus Elite with twice this specific power output (80 bhp per litre). The correct plug for the Lotus would run too cool for the Mini and would soon give misfiring in traffic. The correct Mini plugs would cause pre-ignition in the Lotus.

The plug used in the Mini has a relatively long seating height, whereas the Lotus plug, with more heat to transmit in a given time interval, has a shorter seat height. The gas capacity is the third factor having a marked influence on the temperature of the nose

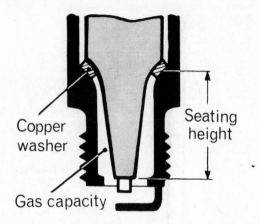

Fig. 2.7   The design parameters in a spark plug.

of the insulator. On a high output engine this capacity is kept small. This reduces the amount of hot gas forced into this cavity during the combustion pressure rise and thus reduces the amount of heat transferred to the insulator.

There is a certain confusion in the nomenclature of heat ranges. In European racing circles, a hot plug is used in a hot engine, i.e. one of high power output per litre. In America, a hot plug is used in a low-powered touring engine and a cold plug in a racing engine. This is all very silly since our aim is to use a plug that is neither too hot or too cold for the particular engine. The author prefers the term *soft* for a plug intended for a *soft* engine and *hard* for a plug used in a *hard-driven* sports/racing engine.

Plugs are an important part of engine tuning. The appearance of the plug, the colour of the insulator nose, the state of the central electrode and the earth electrode(s), the amount and condition of the carbon on the insulator or the points; all these are symptoms that can be interpreted by the experienced mechanic and – with a little guidance – by the DIY tuner. This aspect will be discussed in more detail in Chapter 4.

3
Carburation

Early chauffeurs found that a wide range of mixture strengths would burn in a petrol engine and in the typical automobile made near the end of the century it was left to the chauffeur to adjust, not only the ignition timing, but the mixture strength. With experience, a good chauffeur would make frequent changes to the proportions of air and petrol to suit the operating conditions. A bad chauffeur would sometimes neglect to make these sensitive adjustments, hence the occasional backfires that are a nostalgic memory of early automobilists.

3.1 The mixture requirements

The carburetter in its early form was a very primitive device but it did at least serve to provide a more effective control of the mixture strength than one could expect to be achieved by a typical chauffeur. A chemically correct mixture strength for a typical petrol is about 15 parts (by weight) of air to 1 part of petrol. With such a mixture the correct amount of oxygen is supplied for complete combustion. It is reasonable to suppose this to be the mixture required at all times. Combustion of hydrocarbon fuels is a complex subject, however, and for various reasons beyond the scope of this type of book we must accept that a richer mixture, closer to 13:1, is necessary to produce maximum power.

For obvious reasons one tries to operate the engine on as little fuel as possible at other times. Actual mixture requirements vary from engine to engine. Some designers are more successful than others in the provision of a combustion chamber configuration

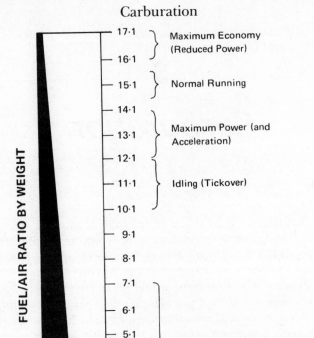

Fig. 3.1 The full range of air/fuel ratios required by an engine.

and an induction system that allows the engine to operate on a lean (weak) mixture that is much closer to the theoretical limit of flammability. The whole range of mixture requirements for a typical modern engine is shown in Fig. 3.1. The mixture requirements are influenced to some extent by the load on the engine. Fig. 3.2 serves to illustrate the influence of load on a typical engine. For maximum power the engine should operate on an air/fuel ratio varying between about 12.5:1 at one-quarter full load and 13:1 at full load. For economical cruising, it is sensible to

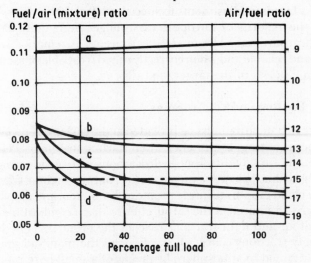

(a) Rich limit of flammability
(b) Maximum power
(c) Maximum economy
(d) Lean limit of flammability
(e) Chemically optimum ratio

Fig. 3.2 The influence of load on the mixture requirements.

operate on the leanest mixture that the engine is able to burn without misfiring or failing to burn completely before the opening of the exhaust valve.

When burning such lean mixtures the throttle opening is always greater than it would be with a chemically correct (stoichometric) mixture, but this is of no consequence at cruising speeds where full throttle is not required. In the particular example of Fig. 3.2 the cruising mixtures vary from 14.2:1 at one-quarter full load to 16:1 at full load.

From the data supplied by the engine manufacturer, the carburetter (or fuel injection equipment) manufacturer attempts to tailor a metering system to match the engine requirements. In the future we hope to see more fuel injection systems in use with mini-computers programmed to match the particular engine requirements. These will be expensive and it will be necessary to leave all tuning adjustments to specialists. For many years, however, the majority of European cars will get along quite happily using carburetters. The modern carburetter is remarkably efficient and reliable and when correctly tuned is capable of metering within 2 or 3% of the target mixture.

3.1.1 Starting and cold-running

A very rich mixture must be provided at the carburetter when an engine is starting, even richer when starting in winter. Petrol is made up of many different hydrocarbons with a wide range of boiling points. When starting in winter, only the lighter fractions of the blend evaporate inside the induction system and a very rich mixture is required at the carburetter so that a combustible mixture of air and light fractions reaches the cylinders. The lower volatility fractions collect on the floor of the manifold and are often drained to atmosphere by means of a small-bore pipe. To start an engine in a typical British winter requires a starting mixture of about 4:1. In colder climates, the starting device or starter carburetter must provide a mixture as rich as 1:1.

During warm-up, the carburetter must still provide a rich mixture but this must be progressively reduced in richness as the engine approaches operating temperature. Older designs of carburetter provide a manual enrichment device for starting and warm-up, the well-known Choke control. With intelligent operation, this can be very economical. Many modern cars have carburetters fitted with enrichment devices that are controlled by a thermostatic sensor. It is not easy to design one of these to give a perfect match to the engine requirements during the entire warm-up process but they are an acceptable compromise for modern traffic conditions when a driver's attention is largely occupied with the problems of driving and cannot be expected to make frequent adjustments to a manual choke.

The use of 'choke' in the context of an enrichment device should be noted. Early cars did actually 'choke' the air supply to the carburetter intake by means of a flap-valve or 'strangler'. In Great Britain, the word choke is also sometimes used to mean the venturi tube which is part of the carburetter metering system. To prevent confusion, this metering choke will be called a 'venturi'.

3.1.2 **Idling**

When an engine is running slowly, the gas velocities through the induction manifold and the inlet valves are so low that mixing of the air and petrol is inadequate. Turbulence inside the cylinders during induction and compression is again far too low to ensure a homogeneous mixture strength inside the combustion chamber at the time of ignition. Not only does the mixture tend to vary from cylinder to cylinder, and from cycle to cycle in the same cylinder, but the mixture strength near the sparking plug is not necessarily the same as that at the opposite side of the chamber. Obviously, any attempt to adjust the idling mixture towards the lean end of the scale will result in spasmodic excessive weakness. The 'splashy' irregular beat of an engine running on a lean idling mixture indicates the failure to achieve good combustion at every firing stroke.

The design of the combustion chamber has an influence on idling behaviour, but all engines require a mixture on the rich side of the chemically correct mixture to give regular idling. The low air velocities also have a deleterious effect on the atomization function of the carburetter. Some designs of variable venturi carburetter do not use special idling circuits to improve atomisation at low air flows, but the majority of carburetters provide an idling circuit, not only to provide a simple means of tuning the idling mixture strength, but most importantly as a method of introducing the metered fuel at a point in the air stream where the velocity is highest, namely at the edge of the almost closed throttle-plate. As the throttle is gradually opened from the idling stop the air velocity across this idling mixture hole falls, and it becomes necessary to introduce an augmented supply from a second hole or series of holes to give what is called 'progression'. An example of this can be seen in Fig. 7.7. The progression orifices (36 in the middle drawing; only one being shown) are above the throttle plate in this down-draught carburetter at idle and are therefore subjected to atmospheric pressure. During idling there is a downward air flow from these progression orifices into the vertical drilled passage to assist in the atomization of the fuel metered into the low-pressure zone below the throttle plate by the idling orifice. As soon as the throttle edge passes above these progression holes, however, the downstream depression begins to draw fuel from these holes too, thus augmenting the total quantity of fuel. The size of the progression orifices is determined on the test-bed by the manufacturer and is beyond our control. Only the idling mixture can be adjusted during tuning. This is performed by adjustment of the mixture screw (part 20 in Fig. 7.7).

3.4.3 Acceleration

When the throttle is opened quickly the depression in the induction manifold drops to a low value and, with no special provision to enrich the mixture, there is a marked tendency to hesitate or misfire from a leaning effect. The reason for the tendency to leanness lies in the condensation of fuel on the manifold walls when the manifold pressure rises. This phenomenon will be discussed more fully later in the chapter. For our present purpose, considering the special case of a sudden change from a high manifold depression to a low one, we must accept that the manifold walls with a high depression (16–18 inches of mercury) and an engine at working temperature will be perfectly dry. When the throttle is snapped open, however, the rise in manifold pressure (2–4 inches of mercury) causes the high boiling point fraction of the petrol to condense on the manifold walls. Unless an additional amount of petrol is supplied by some enrichment device the mixture reaching the cylinders must be too lean. The popular solution to this problem is the provision of an accelerator pump that injects a metered amount of extra fuel during acceleration.

3.2 The basic carburetter

A carburetter can be described as an instrument capable of mixing air and a liquid hydrocarbon fuel (normally petrol) in the required metered proportions to ensure ready ignition and efficient combustion. A method of controlling the power output of an engine is essential. For convenience this is incorporated in the carburettor body. It takes the form of a throttle-plate which can be rotated from a near-closed position to a full-throttle position when an almost unobstructed passage is presented to the airstream.

The basic carburetter always uses the depression created by a shaped restriction or venturi to act as the primary system for inducing fuel into the airstream. Carburetters can be divided for convenience into two types, *fixed venturi* and *variable venturi,* the SU being the best known of the latter type.

3.2.1 The fixed venturi carburetter

This common type of carburetter presents a tapered restriction, or venturi, to the airflow. This speeds up the air velocity at the narrow neck and produces a depression, i.e. a drop in pressure. Fig. 3.3 is a simplified drawing of the metering system. A constant

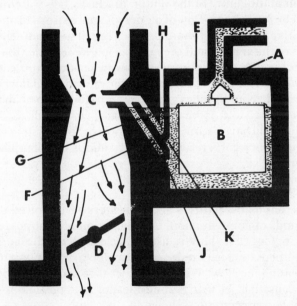

A	=	Needle Valve	F	= Channel from Main Jet
B	=	Float	G	= Channel from Compensating Jet
C	=	Choke Tube	H	= Reserve Well (Riser Tube)
D	=	Throttle	J	= Main Jet
E	=	Float Chamber Vent	K	= Compensating Jet

Fig. 3.3   The basic fixed venturi carburetter (original Zenith).

head of petrol is maintained in the float chamber in the manner we have all seen in domestic water cisterns. The pressure above the petrol in the float chamber is atmospheric, or slightly below, since in modern installations the float chamber is vented to the zone between the air cleaner and the carburetter in the interests of cleanliness.

Petrol is fed to the centre of the venturi by a simple duct. Since the exit of the duct is at a slightly higher level than that in the float chamber a certain minimum airflow is necessary to induce sufficient suction head to cause petrol to flow into the venturi. As the air velocity increases the depression also increases and the amount of fuel sucked into the venturi increases. Unfortunately, the system will not maintain a constant air to fuel ratio. The flow of fuel increases in proportion to air *velocity*. Since the density of the air at the centre of the venturi gradually falls with increase of air velocity the mass-flow of air does not increase in proportion to the increase in mass-flow of petrol. Thus, if we choose the dimensions of the venturi and the main jet to give us an air to fuel ratio of 15:1 at 25% of full throttle air rate, when the throttle is opened further to give 50% of full throttle air rate, the fuel flow will have increased about 2½ times giving an air to fuel ratio of about 12:1. With a four-fold increase in air rate, i.e. to full throttle, the mixture will have become excessively rich. Some system to correct this tendency is therefore necessary and is usually called 'compensation'.

Mixture compensation

Different manufacturers of carburetters have evolved their own methods of compensation. Compensation is illustrated graphically in Fig. 3.4. This is usually achieved by the addition of a compensating jet designed to meter a leaner mixture with increasing air flow. With a main jet that becomes richer and a compensating jet that becomes leaner it is possible to strike a suitable compromise as shown in the graph.

Zenith compensation

Older Zenith carburetters used the system shown diagrammatically in Fig. 3.5. The compensating jet feeds into a capacity or reserve well that is open to atmospheric pressure (or to the float

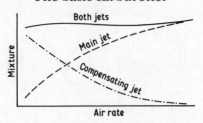

Fig. 3.4 The Zenith principle of compensation.

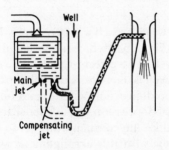

Fig. 3.5 Zenith compensation.

chamber vent passage). Fuel flow from the compensating jet is constant, since it is governed by the head of petrol in the well and the size of the compensating jet. As the air flow increases the mixture contribution from the compensating system must therefore become progressively leaner. A secondary effect of this compensating system is that air bleeds into the fuel forming an aerated feed to the venturi, i.e. a finely dispersed aerated mixture. The carburetter manufacturers have always claimed that this emulsification is beneficial in that it improves the air/fuel mixing function of the carburetter. Even so the air bleed introduced by the compensating system is unavoidable.

Solex compensation

The system used by Solex, Weber and other makers of fixed venturi carburetters is a subtle modification of the Zenith compensating jet. Only one jet, the main jet, is used. A variable air bleed gives the required compensation. The method can readily

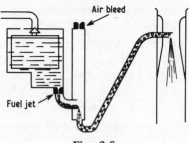

Fig. 3.6.

be understood if we approach it in two stages. The first stage is the system shown in Fig. 3.6 in which an air bleed is added to the top of the capacity well. The effect of placing a restriction to the air flowing into the well is to create a partial depression in this zone. The extent of the depression will depend on the size of the air bleed. This partial depression will influence the flow of fuel from the main jet. Unlike the compensating jet in the Zenith carbur-etter the fuel flow with this arrangement would increase with increase in air flow. At first glance, it would appear that we have now devised a metering system that will maintain a fairly constant mixture strength over a wide range of flows. Unfortunately, the same physical laws apply to the flow of air through the air bleed as to the flow through the venturi. On a mass-flow basis the mixture would become richer with increasing air flow.

The Solex solution is to introduce a second air bleed that increases in size with increase in air flow. The principle of the Solex emulsion tube is shown diagrammatically in Fig. 3.7. The emulsion tube is an extension of the air bleed (correction jet). This tube has several rings of drilled holes at different levels. At low air flows through the venturi there is negligible pressure difference between the inside and outside of the tube and the only holes above the fuel level are those in the top ring. As the air consumption of the engine increases so the venturi depression increases and with an increase in venturi depression a greater air flow is induced through the emulsion tube. This increased air flow through the holes in the emulsion tube tends to increase the pressure drop across the tube. This drops the level on the outside relative to the inside, thus exposing another set of holes. In effect,

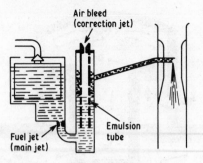

Fig. 3.7 Solex compensation.

then, the air correction jet is not exposed to the full venturi depression; the main jet is. By careful selection of jet sizes and emulsion tube hole sizes a relatively constant (or rising and falling) air to fuel ratio can be supplied to suit the engine requirements. A secondary effect of the emulsion tube is the aeration (or emulsion) created by the air bubbling through the small holes in the tube. This is not strictly an emulsion as the term is used by chemists, but the term has always been used by carburetter people and the aeration of the mixture entering the venturi must assist in the formation of a finely divided droplet spray.

3.2.2 The variable venturi carburetter

The working principle of the variable venturi carburetter is shown in Fig. 3.8. In the SU carburetter, the effective area of the venturi is varied by the rise and fall of a piston, P, under the suction created in an upper chamber, R. In the Stromberg design, a synthetic rubber diaphragm controls the throat area, but the working principle is identical. In the simple SU layout of Fig. 3.8, fuel is supplied to the single jet at a constant head by means of a conventional float chamber. A tapered needle mounted in the base of a moveable piston increases the effective area of this jet as the piston rises. This venturi piston is integral with a larger diameter control piston which moves freely inside the air valve chamber. The upper side of this larger piston is subjected to the depression existing in the zone between the throttle plate and the venturi piston. By design, this is a depression of about 8 inches of

P = Piston (Air Valve)
Q = Carburetter Barrel
R = Air Valve Chamber
S = Effective Choke Area

Fig. 3.8   The basic variable venturi carburetter (SU).

water. The annular space between the two pistons is maintained at atmospheric pressure by means of a balance hole.

At idling engine speed, the base of the small piston rests on two small protuberances, thus leaving a small air passage for the idling air flow. With the variable fuel jet adjusted to the right height by careful tuning, as will be described later, the correct idling mixture will be given when this fuel mixes with the air passing below the piston. As the throttle is opened gradually the air flow through the passage will increase and the depression downstream of the small piston (and above the large one) will eventually exert an upward force that exactly balances the weight of the piston assembly. Any further increase in throttle opening will cause the piston assembly to rise. The piston will rise until the increase in venturi area restores the depression of 8 inches of water to the zone upstream of the throttle plate. By selecting a needle with a suitably tapered profile (and SU make more than 500 differing profiles) the mixture strength can be maintained constant at all times or varied according to any chosen programme. This is the basis of the SU and Stromberg metering systems. Certain refinements are necessary to provide starting enrichment and economical cruising and these will be described in Chapter 6.

3.3 **Mixture formation and distribution**

In theory, the work of the carburetter designer stops downstream of the throttle plate. In practice, there is usually co-operation between the development departments at the engine works and the carburetter manufacturer in the design of the induction manifold. Perfect mixture formation and distribution would be said to occur if the mixture arriving at each cylinder in turn had an identical air to fuel ratio and an identical boiling point range. This seldom happens. Even in a warm climate very little evaporation takes place in the actual carburetter. The amount of fuel evaporated inside the manifold depends largely upon the manifold depression, since the boiling point of any liquid is always lower at a lower ambient pressure. A typical petrol blend is a mixture of hydrocarbons with boiling points (at atmospheric pressure) ranging from a small percentage of butane with a boiling point of 40°C to the least volatile fraction with a boiling point of about 220°C. The high volatility fraction is necessary for cold starting.

With a high manifold depression, i.e. when starting from cold or when idling, all the butane and a certain percentage of the more volatile fractions will flash off immediately into vapour as the wet mixture passes the throttle edge. Depending upon manifold design and the amount of heat supplied to it, more of the middle boiling point range will evaporate inside the manifold. The percentage of the total mixture entering the cylinder in a liquid state is therefore dependent upon many factors, the most important of which are:

(a) Manifold depression.
(b) Amount of external heat applied to the manifold.
(c) Atmospheric temperature.

In general we would expect to find a portion of the mixture still in the form of liquid droplets as it enters the cylinder. During the induction and compression strokes, heat is picked up from the hot surfaces in the cylinder head, particularly from the hot exhaust valve head and from admixing with the hot gases retained from the previous cycle. With an engine at normal operating temperature all the mixture is in the form of petroleum vapour and air by the time the ignition occurs. With a cold engine, however, a large proportion of the petrol is still in the liquid state. The ability to fire such a wet mixture is a measure of the effectiveness of modern ignition technology.

3.3.1 Manifold design

At light loads, i.e. when cruising at about 40–50 mph about one-third of the mixture passing through the induction manifold is still in the liquid state. At higher power outputs and lower manifold vacuums, only about one-half of the mixture has been evaporated by the time it enters the cylinder. This dual-phase condition of the mixture has always presented a difficult problem to engine designers and, at best, the induction manifold never gives perfect distribution.

For good power output the passages in the manifold should be proportioned to give as little resistance to the passage of the mixture as possible. This calls for generous cross-sections and a smooth blending of the branches and the main manifold duct. Good distribution of the mixture demands passages that create a high degree of turbulence, especially when the mixture leaves the main duct to enter a branch. High turbulence is necessary to keep the liquid droplets in suspension. Even so at a low manifold vacuum (wide open throttle), especially with a low engine speed as when climbing a hill, it is difficult to prevent a stream of liquid from flowing along the floor of the manifold and preferentially entering the branch to a particular cylinder. A hot-spot is sometimes provided. This is a sump or well at the lowest point of the manifold which is heated by contact with the exhaust manifold or by the provision of a water jacket. The exhaust heated hot-spot is most effective when the engine is cold since the low volatility fuel collects in this well where it extracts sufficient latent heat to evaporate fairly quickly.

Manifold design is therefore seen to be a matter for compromise. In a racing engine, the designer concentrates on achieving a low pressure drop. By using separate carburetters or separate venturis to each cylinder he can overcome the problems of poor distribution. For a touring engine with only one, or at the most two carburetters, the designer sacrifices a little power and designs the manifold to promote a modest degree of turbulence, thus breaking up the larger droplets and giving a fairly even mixture to all cylinders. Typical mixture scatter at full power on a 4-cylinder would be $\pm 2\%$ variation in the air to fuel ratio. At idle this could increase to $\pm 6\%$. This represents a total variation of about $1\frac{1}{2}$ ratios from the leanest to the richest cylinder. A mean of 12 to 1 would vary from 11.3 to 12.7. This is acceptable; anything wider would make it impossible to tune the carburetter to give a regular clean idle.

4

Testing

Well-known systems of professional tuning, as used at Crypton and Sun Tuning Stations, always start with a series of tests. From the results of these tests the tuner is able to diagnose the condition of the engine from all aspects. His diagnosis may suggest that minor or even major maintenance work is necessary before tuning work can be started. Cynical readers will no doubt suggest that some tuning stations are liable to find 'necessary maintenance' in order to run up the size of the bill. The author does not believe that many garages do rob the customer in this way, but those who disagree have the solution in the pages of this book. They can always tune their own engines!

4.1 Test no. 1: the road test

4.1.1 Acceleration

One should always begin with a Road Test for one's own car. If an 'identical model' surges past you on a hill and convinces you that your car has lost its tune you should have a 'yard-stick', a simple road test to be carried out on your home ground to establish if your car has indeed lost its tune. One can sometimes be deceived. Perhaps the 'identical model' had a larger engine or had been 'souped-up' to produce more power (usually at the expense of economy).

Thirty years ago, when roads were less congested, the writer had a measured quarter-mile on a quiet road which he used to measure both the standing-start time for the quarter-mile, at the same time noting the terminal speed. From experience he found the terminal speed to be more consistent as a measure of the state of tune than the elapsed time taken by stopwatch. A modern high performance saloon can reach 80, 90 or even 100 mph at the end of a quarter-mile. It would be very foolish to use such a test method today. A fairly steep main road hill, however, could be used today without any danger. The test length should be short enough to keep the terminal speed below about 50 mph.

Proceed as follows:

(1) Park at the kerb with the front of the car in line with a suitable starting mark such as a lamp standard.
(2) Having chosen a final mark, such as the second or third lamp standard further up the hill, observe the state of the traffic and chose a safe time to start the test.
(3) Accelerate through the gears, with one eye on the speedometer, taking care to change gears at the same speed in every test.
(4) Note the speed as the front of the car passes the final mark.

This test is a good measure of the general state of tune of the engine. It tests power over a range of speeds. It shows up general ignition faults, such as plug misfiring, incorrect ignition timing or incorrect distributor advance characteristics. It shows carburation weaknesses and any mechanical failings such as badly seated valves or broken piston rings.

If the Reference Terminal Speed has been well established for the correctly tuned car one should begin to suspect that something is no longer functioning efficiently if the actual terminal speed has fallen by 10% or more. A drop of only 5–7% could quite easily be caused by a combination of unfavourable atmospheric conditions and variations in driving technique.

4.1.2 Timed tests

If a stop watch is available and an assistant can be enlisted to operate it we can establish quite accurate timed reference figures. Two good tests for a typical car with a 4-speed gearbox would be:

(1) The time to accelerate between 30 and 50 mph in top gear.

(2) The time to accelerate between 30 and 50 mph in third gear.

In this way we measure both low speed and medium speed torque. We are not interested in high speed torque. With certain ignition or carburation irregularities it is possible for tune to be lost at one end of the speed range and unaffected at the other. A test of this type should reveal such changes.

4.1.3 Fuel consumption

If the reader has not already carried out reliable fuel consumption measurements on his car he should check the mpg (or km/l) over a well established route before tuning the engine. A reasonable method requiring no special equipment is to measure the overall consumption over a distance of at least 500 miles. The tank should be filled to the top at the start and refilled again at the end. Obviously, the test conditions should represent the typical traffic pattern experienced in day to day driving by the car. If, for example, 75% of the mileage is on a regular commuter run, 10% local shopping and 15% pleasure motoring, this pattern should be repeated as closely as possible when testing before and after tuning.

4.2 Test no. 2: plug condition

To learn anything from the appearance of the spark plugs it is essential to read their message immediately after the engine has been running under full load. This, then, falls naturally into order as the second test, since we can follow the acceleration tests by an immediate return to our garage. The car should be driven fairly briskly up to the garage entrance, then the gear lever should be put into neutral and the ignition immediately switched off. If the engine is allowed to idle before the plugs are examined the colour evidence will be destroyed.

The reader can check in his Driver's Manual or Workshop Handbook that the correct plugs are in use. His local garage will have the information if he does not possess a Handbook. The author once bought a Riley with a tendency to pre-ignite. A former owner had fitted the wrong plugs. Plug no. 3 (on the upper row) in Fig. 4.1 shows the eventual fate of such a plug. The plug is obviously destroyed. What is less obvious is that prolonged pre-ignition or heavy detonation will cause engine overheating. This could lead to piston seizure or burned exhaust valves. The broken insulator in the second plug on the upper row suggests that a clumsy mechanic was at work when the gap was last set. Care should always be taken to avoid putting pressure on the central electrode when opening the gap. The ceramic insulator is very brittle. Plug no. 4 indicates an over-rich mixture. If, however, the black deposit is not dry, but sticky and oily, this particular cylinder is suffering from excessive oil consumption. The oil could be passing the piston rings, or less frequently, could be entering the combustion chamber down the inlet valve guides during the induction stroke. Chronic cases of oiled plugs call for immediate maintenance work.

If we find that all plugs are as shown in the sixth figure, with a light brownish-grey colour, we should be happy. No combustion problems are indicated and the mixture strength under load must be close to the optimum.

4.3 **Test no. 3: cylinder compression pressures**

Oh for the days when we could test compression on the starting handle! It was not very accurate, but it was certainly quick and easy. No matter; it is now a matter of History that we allowed the car makers to rob us of this very useful accessory. Today, we need a special pressure gauge to measure compression pressures. It is special in that a special adaptor is required to make a pressure-tight seal at the plug-hole. One pattern uses a rigid connecting tube and a tapered rubber bung that can be pressed firmly into the plug-hole. This type is quick to use, but the rubber bung eventually becomes lacerated by the sharp cutting edge of the first thread at the top of the hole. The gauge using a screwed adaptor is obviously slower in use but is more reliable. If the car is an old one we should not assume that the compressions measured, even after tuning, are as good as those when new. The Workshop Manual should be consulted. This will tell us if the mechanical state of the engine has deteriorated. The Manual will usually state whether the throttle should be fully open or closed

Electrodes burnt away due to wrong heat value or chronic pre-ignition

Plug in sound condition with light greyish brown deposits

Broken porcelain insulation due to bent central electrode

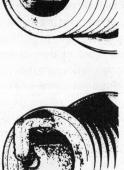

Mild white deposits and electrode burnt indicating too weak a fuel mixture

White deposits and damaged porcelain insulation indicating overheating

Excessive black deposits caused by over-rich mixture or wrong heat value

Fig. 4.1 Reading the plugs.

when taking the compression pressures. Naturally, even at the low engine speeds given by starter rotation, slightly higher figures are given with the open throttle option.

The engine should be at working temperature before carrying out the tests.

4.3.1 Dry compressions

The compression gauge is fitted to each plug hole in turn, all plugs having been removed first to ease the load on the starter. When an auxiliary starter button is not provided in the engine compartment it is useful to have an assistant in the driver's seat to operate the starter button (ignition switch). With the adaptor fitted to the first plug-hole the starter is operated. Since all the other cylinders have their combustion chambers vented to atmosphere, the engine revolutions can readily be counted as all cylinders but one produce a puffing noise from the plug holes. The number of revolutions are counted and the starter switch released after eight revolutions. A maximum reading is usually reached after five or six revolutions. The gauge is fitted with a non-return valve. The maximum reading is therefore retained until the release button is pressed. This first reading is recorded, the gauge re-set to zero and the operation repeated until all dry compressions have been recorded.

If one or more compressions is more than 15% below the average, a wet compression test should be made.

4.3.2 Wet compressions

The purpose of this test is to provide a temporary seal around the pistons with a film of oil. Should a much higher reading be obtained on any cylinder by this wet test, broken or worn piston rings are indicated.

Approximately one tablespoonful of light engine oil (SAE Grade 20 or 1OW/40) should be poured into each plug hole before measuring the compression. Pouring in the oil is not always easy. The professional method is to use a pump-type oil can, first measuring how many pump strokes represent a tablespoonful. Immediately after adding the oil to no. 1 cylinder the *wet* compression reading should be taken and recorded. A maximum reading should be achieved after about six revolutions. The procedure should be repeated on all cylinders.

If the valves are all seating well, with no tendency to stick in the guides, we would expect the compression pressure on any suspect cylinder to be temporarily restored by the film of oil around the top ring. If, for example, no. 3 cylinder is suffering from broken piston rings or, unhappily, a scored cylinder bore we would expect a set of readings as indicated below:

Cylinder no.	1	2	3	4
Dry compression (psi)	160	155	127	161
Wet compression (psi)	173	169	168	172

If, however, the piston rings and cylinder bore on No 3 cylinder are in excellent condition, but the exhaust valve seat is no longer gas-tight, we would expect the following pattern to be shown:

Cylinder no.	1	2	3	4
Dry compression (psi)	160	155	127	161
Wet compression (psi)	173	169	139	172

There is one other case to consider. If two adjacent cylinders indicate abnormally low compression pressures and injecting oil for a wet test does not bring the pressures into line with the other cylinders the cause could be a leakage of gas between the two cylinders across the head gasket. Examination of the coolant in the radiator header tank or in the expansion tank could show a film of oil on the surface, further evidence suggesting a leaking head gasket.

One would naturally want to repeat the compression tests before proceeding further, but there would be little point in attempting to tune an engine with worn or broken piston rings, badly seated valves or a leaking head gasket. These mechanical defects would be corrected, either by a visit to your local garage or, if the reader is bold enough, with your own unaided efforts, a good set of tools and the Workshop Handbook.

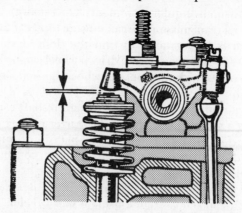

Fig. 4.2   Measuring valve clearances on push-rod engine. Rocker clearance is measured at point shown.

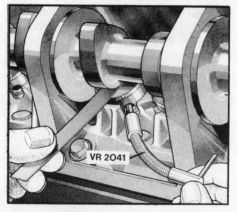

Fig. 4.3   Measuring valve clearances on o.h.c. engine (Vauxhall).

4.4 **Test no. 4: valve clearances**

On the typical push-rod o.h.v. engine, checking the valve clearances is a simple operation. The plugs should remain out during this operation, since we want as little resistance as possible to engine rotation. Without the assistance of the aforementioned starting handle we have a choice of methods to make the engine turn over by small angular increments. On a smooth garage floor and a manual transmission we can engage top gear and push the car backwards or forwards to find the desired valve opening position. On rough ground one driven wheel can be jacked up to clear the ground, with the opposite wheel chocked, and the wheel turned with top gear engaged.

With an automatic transmission the crankshaft can be rotated by pushing a screwdriver against the starter ring gear. Access to the ring gear is usually through a small plate or grommet fitted at the top of the converter housing. If in doubt, consult the Workshop Handbook.

Fig. 4.2 shows where the valve clearance is measured on a typical push-rod o.h.v. engine using a set of feeler gauges. The same operation is shown in Fig. 4.3 carried out on a single overhead camshaft engine. Valve clearances are usually measured with a cold engine, but some makers specify clearances with a hot engine. Since valve clearances must be measured with the particular valve fully closed it is necessary to rotate the engine by one of the suggested methods after each measurement. A useful guide on a 4-cylinder in-line engine, for which the firing order is invariably 1–3–4–2, is to remember the number 9. This will be seen in the table below.

Measure no. 1 valve clearance with no. 8 valve fully open
Measure no. 3 valve clearance with no. 6 valve fully open
Measure no. 5 valve clearance with no. 4 valve fully open
Measure no. 2 valve clearance with no. 7 valve fully open
Measure no. 8 valve clearance with no. 1 valve fully open
Measure no. 6 valve clearance with no. 3 valve fully open
Measure no. 4 valve clearance with no. 5 valve fully open
Measure no. 7 valve clearance with no. 2 valve fully open

The valves are numbered from the 'front end' of the engine, i.e. the opposite end to the flywheel end.

With a 6-cylinder engine with the normal firing order of 1–5–3–6–2–4, the following procedure will save time and energy. Again counting the valves from the front, nos. 1 to 12, with no regard to their duty, rotate the engine in the normal direction until no. 1 valve is fully open. Valve clearances of nos. 3, 7 and 11 can now be measured. Rotate again until no. 4 is fully open. The clearances of nos. 5, 10 and 12 can now be measured. Similarly, when no. 2 is open, no. 6 can be measured and with no. 8 just open, the final clearance, no. 9 can be measured.

All valves will probably be found to have some clearance. With the passage of time exhaust valves have a tendency to sink into their seats. This reduces their clearance. Any exhaust valve with no clearance will probably fail in the very near future. Incipient valve failure in any cylinder could already have been indicated by the compression readings.

4.5 Test no. 5: the fuel system

4.5.1 Carburetter(s)

Carburetter manufacturers have a habit of complaining that their products are often blamed for engine malfunctions that should be attributed to the ignition system and sometimes to mechanical defects. In the writer's experience this is justified. Carburetters seldom give any trouble. It therefore follows that we would be wasting valuable time if we checked the state of tune of the carburetter before looking at the ignition system. At this stage, therefore, we only recommend an external examination of the carburetter(s). Check that the throttle controls are working freely and that they return to the idle stop every time. With Bowden cables look for signs of dirt or rust where the inner cables enter the outer. If the carburetter is controlled by a rod and lever system, check that all pivot bearings are free and well lubricated. When a manual choke is fitted, open and close this control and check that it moves freely and returns to the fully off position at the carburetter end. Finally check for leaks. Leaks may not be apparent if the car has been standing for a while, since the level in the float chamber will have dropped while standing. With an SU electric pump a leak is easily detected by switching on the ignition. After the initial ticking noise of the pump action as the float bowl level is restored (this should only be from evaporation), there should be complete silence. If the pump continues to tick at

regular intervals, even with an interval of several seconds between ticks, there must be a leak somewhere. It could be a flooding float chamber, or a leaking connection at the carburetter or at
the fuel pump, or even at some point between them. With a
mechanical pump one should first run the engine for about a
minute, then switch off and examine the carburetter, feed pipe
and fuel pump for any signs of wetness. Fuel leaks will obviously
drain to the lowest point and can often be found by running the
hand over the base of the carburetter and the pump. Pipe connections should be checked with a spanner. Where carburetter
flooding is suspected the condition of the needle and seat should
be checked after removing the float bowl. The repair kits, available from the carburetter stockists or the local car agents, usually
contain a replacement needle and seat.

4.5.2 Crankcase emission control valve

The earliest automobile anti-pollution law passed by the all-
powerful State of California made automobile manufacturers all
over the world fit a simple emission control system for the gases
that had previously been permitted to escape from the crankcase
breather into the atmosphere. After 1965, many cars sold outside
California were fitted with a pipe that fed the crankcase gases into
the induction system. From our point of view this was an economy
measure, even on a car with good cylinder bores and good piston
rings, since all cars pass a certain amount of unburned mixture
into the crankcase. Surprisingly, an analysis of the vapour in the
crankcase shows it to be largely unburned petrol mixture with a
small percentage of oil vapour. A non-return valve is always fitted
at some point in the pipe between the crankcase and the induction
system. This prevents the passage of flame from an induction
backfire ever reaching the crankcase. There is a tendency for this
valve to become coated with a gummy deposit from condensed oil
vapours. With neglect the valve can become inoperative. It is a
simple matter to check that the valve is working freely and, if
choked with a mixture of emulsified oil and water, it can easily be
washed clean with a cupful of petrol.

4.5.3 **Fuel pump**

Two tests can be carried out on the petrol pump, a delivery rate test and a static pressure test.

Delivery rate

Testing a mechanical pump always requires the addition of a T-piece in the line between the pump and the carburetter inlet since the engine must be run to create fuel pressure and flow. An electric pump will operate as soon as the ignition is switched on. A T-piece is therefore not required.

With a mechanical pump the fuel line to the carburetter is first disconnected. Three lengths of suitable rubber hose are then connected to each end of the T-piece. Connections are made to the pump delivery and the carburetter feed-line. The third length of hose is inserted into a suitable graduated jar, or any beaker or other vessel of known capacity. With an electric pump only one connection is necessary, from the pump outlet into the measuring vessel.

With a mechanical pump the engine is started and run at a speed of about 1000 rpm. With the electric pump only the ignition need be switched on. The time to half-fill the measuring vessel should be taken. No great accuracy is necessary and a wrist-watch with a second hand can be used to time the rate. With the larger pumps, such as the SU model AUF400, only about 10 to 15 seconds of operation will suffice, since there is usually no space available to accommodate a large vessel. With large pumps the assistance of a helper to start and stop the engine or to switch the ignition on and off is advisable. A flood of petrol on the garage floor is both wasteful and dangerous.

Delivery rates of SU pumps

Model	Normal delivery, l/min
AUF 200 (small single pump)	0·5
AUF 300 (medium single pump)	1·1
AUF 400 (dual pump)	2·2
AUF 800 (mechanical pump)	0·8

The capacities of other makes and models of pump are usually quoted in the car Workshop Manual.

Static pressure

For this test, a pressure gauge reading up to 5 psi (0.35 kg/cm^2) is necessary. Command (Narco National Ltd.) market combined vacuum gauge and fuel pump tester. This is obtainable at most accessory shops.

With mechanical pumps the T-piece is still required, the third branch this time being fitted to the pressure gauge. The gauge can be fitted directly to the pump outlet in the case of an electric pump. The pressure from the mechanical pump should be measured at several speeds from idling up to 4000 to 5000 rpm. A typical mechanical pump will give a pressure varying between 3 psi (0.21 kg/cm^2) and 4 psi (0.28 kg/cm^2). The majority of SU electric pumps should give a static pressure of 2.7 psi (0.19 kg/cm^2). In some installations, however, the car manufacturer has requested a higher pressure from the larger models of SU pump and these are designed to give a static pressure of 3.8. psi (0.26 kg/cm^2).

When a pump fails to meet its specification for pressure or delivery rate one should first check that the pump filter is clean. A failing mechanical pump is usually caused by a leaking diaphragm or a worn suction or delivery valve. Replacement kits may be obtained from the pump stockists or the car agents. They are relatively easy to fit. Electric pumps are not as easy to service, but an exchange pump service is available at most agents.

5

Tuning

5.1. **A chore or a hobby?**

Tuning for economy can make an interesting hobby. With some car owners it can even become an obsession. We have assumed that the reader of this book is a typical motorist trying to keep his car in a good state of tune in a fairly painless way. Our aim is to get within 2 or 3% of *optimum economy tune*. A professional tuner with thousands of pounds-worth of sophisticated equipment at his command would be more effective, but his services are only necessary if the reader is too busy, too ham-fisted, too lazy or even rich.

The six operations described below can be carried out any Saturday morning, with a break for coffee. With the exception of the first operation it is advisable to repeat them every 6000 miles. If pressed for time one could extend this period to 8000 miles. The first operation, the setting of valve clearances, could be omitted at alternate tune-ups.

5.2 **Operation no. 1: valve clearance (12 000 miles)**

On a push-rod o.h.v. engine this is a fairly easy operation. The clearance is adjusted in the manner shown in Fig. 5.1. The lock nut is slackened, the gap opened or closed as required by turning the screw-driver slot in the top of the rocker ball-end. With the correct clearance indicated by the feeler gauges (the right combination of feeler thicknesses will just slide in the gap – with one extra thou, the feelers are too tight), the lock nut is re-tightened.

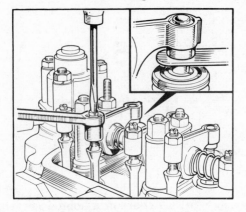

Fig. 5.1 Setting valve clearance on a push-rod engine.

As the lock nut is re-tightened the screw-driver should be held firmly in the slot to prevent movement. After locking up, it is advisable to re-check the clearance, since a slight movement of the ball-end when tightening could reduce the clearance. The valve must be closed, of course, when setting the clearance (see Chapter 4).

Overhead designs fall into two main types. The first type uses a pivoted follower, rather similar to the rocker used on the push-rod o.h.v. engine. Adjustment of the clearance is carried out by a method similar to that used for the push-rod engine. The second type is that represented by the classic Jaguar double overhead camshaft engine which uses inverted bucket tappets acting directly on the ends of the valve stems to transmit the cam lift to the valves. The more modern single overhead camshaft layout, as used on the V-12 Jaguar engine is shown in cross-section in Fig. 5.2. Adjustment of valve clearances is a fairly tedious operation on engines of this type. Fortunately, they hold clearance for long periods, since engines of this type are usually fitted with valves and valve seats of superior material. Adjustment of valve clearances involves the replacement of the small hardened disc that fits between the valve stem tip and the base of the bucket tappet by one of different thickness. Using the clearance readings noted down in test no. 4 of the last chapter we must make a visit to the

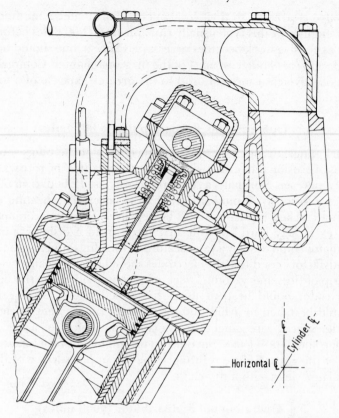

Horizontal ℄
Cylinder ℄

Fig. 5.2 Single o.h.c. head (Jaguar). Clearance is measured between back of cam and bucket tappet.

spares department of the local dealer to obtain the necessary discs of the thicknesses that will restore the correct clearances as stated in the Handbook. With luck, a few of the existing discs can be re-used in a different position. To complete the operation, the camshaft(s) must be removed to give access to the bucket tappets. The Maker's Handbook will give precise information on this operation. It is essential not to disturb the valve timing and those with no great mechanical aptitude are well advised, particularly in the case of double overhead camshaft engines, to trust the setting of valve clearances to their local agents. Fortunately, one finds that engines of this type usually run at least 20 000 miles before this expensive work becomes necessary. Even so, one should not neglect to carry out Test no. 4 of the previous chapter. Complete loss of valve clearance can lead to the greater expense of a top overhaul.

5.3 **Operation no. 2: spark plugs (6 000 miles)**

Spark plugs are not expensive. If one or two of the plugs show signs of erosion the reader will be rather niggardly if he refuses to buy a new set. Normally, a set of plugs will last 12 000 to 15 000 miles. Obviously, a plug suffering from severe overheating or with a cracked insulator would be replaced. Insulator ceramics are extremely hard 'and one tiny piece from a disintegrating insulator could destroy a valve seating. If the plugs appear to be in a satisfactory condition the electrodes should first be cleaned with a strip of fine emery cloth. To set the recommended gap the side electrode should be bent using fine-nosed pliers. Great care should be taken to avoid putting pressure on the central elec- trode. Special wire gauges are available. Champion sell a special gauge and gapping tool, but one can use engineer's feeler gauges to set the gaps. Before refitting, wipe the upper surface of the insulator clean with a dry cloth.

5.4 **Operation no. 3: distributor (6000 miles)**

Contact breaker gaps can increase in two ways. The plastic heel in contact with the cam face, despite the availability of improved materials, still suffers gradual wear. The actual contacts lose metal by arcing. Metal is also transferred from one contact to the other, forming a crater on one point and a corresponding mound on the other. If the points are badly burned or the crater is large it is advisable to fit a new contact set. For less extensive wear the mound should be removed by the use of a fine carburundum stick. This work can be facilitated when the distributor is not very accessible, by removing the distributor from the engine. The distributor cap is first removed by pulling back the spring clips. These are quite strong and a screw-driver can be used to provide leverage. Next disconnect the vacuum advance pipe, if fitted (see

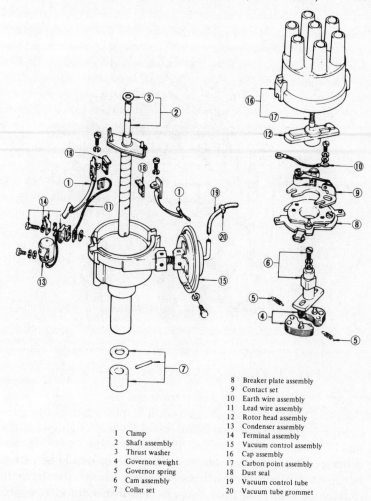

1 Clamp
2 Shaft assembly
3 Thrust washer
4 Governor weight
5 Governor spring
6 Cam assembly
7 Collar set

8   Breaker plate assembly
9   Contact set
10 Earth wire assembly
11 Lead wire assembly
12 Rotor head assembly
13 Condenser assembly
14 Terminal assembly
15 Vacuum control assembly
16 Cap assembly
17 Carbon point assembly
18 Dust seal
19 Vacuum control tube
20 Vacuum tube grommet

Fig. 5.3   Exploded view of distributor (Hitachi model D606-52).

Fig. 5.3), and finally disconnect the LT lead. Before withdrawing
the distributor body mark the position with a metal scriber,
making one line on the base of the distributor, or its clamping
plate, and another on the base plate. On most designs the distributor body can be withdrawn after the bolt(s) securing the

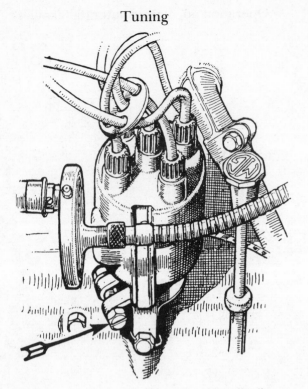

Fig. 5.4 Clamping bolt on a typical Lucas distributor.

clamping plate have been removed. The pinch-bolt holding the clamping plate to the distributor should remain tightened. The pinch-bolt on a typical Lucas distributor is shown in Fig. 5.4. When re-fitting the distributor the drive shaft should be turned by the rotor arm until the offset driving key engages the slot in the distributor shaft. On some engines the distributor is accessible and work on the contacts can be carried out *in situ*.

In fitting a new contact set it is essential to replace the new components and attach the leads in the correct order and position. If the Maker's Handbook gives no clear guidance one should note the relative position of the components when removing the old contacts. Before fitting, the protective grease should be removed from the contact faces with a petrol-soaked rag.

A typical example is the Lucas 25D4 as fitted to the British Leyland 1100 cc push-rod engine (see Fig. 5.5). The contact set is removed as follows:

(a) Remove the nut (8), lift the top insulating bush and both leads from the stud.
(b) Remove the screw (6) with its spring and plain washers.
(c) Remove the old contact set.
(d) Lubricate the pivot post (2) sparingly and check that the insulating bush (9) is correctly positioned below the spring loop.
(e) Place the one-piece contact set on the distributor plate and lightly tighten the screw (6).
(f) Locate the lead terminals around the insulating bush so that they make contact with the spring and tighten the nut (8).

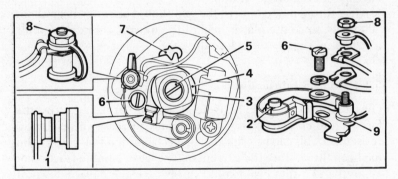

Fig. 5.5   Fitting a new contact set to a Lucas 25D4 distributor.

5.4.1 **Setting the contact gap**

After fitting a new contact set or re-facing an old set the contacts should be checked for misalignment (see Fig. 5.6) and then set to the recommended gap in the manner described in the manual. We can take the Lucas 25D4 as a typical British example. The crankshaft is first rotated until the contacts are fully open. Screw (6) in Fig. 5.5 should be slackened and a screw-driver inserted at (7) to engage the two V-shaped notches. Turning the screw-

Tuning

(a)   (b)

Fig. 5.6 Contact misalignment.

driver clockwise decreases the gap; anti-clockwise increases. The
gap can be checked by means of feeler gauges. When the correct
gap has been obtained screw (6) should be fully tightened. A
second check is advisable after once more rotating the engine and
re-measuring the gap with the contacts fully open. When a car is
reaching 'middle age' the distributor bearings become worn and
the same gap is no longer given at each cam lobe. When this
difference becomes excessive a replacement distributor is the
only way to regain good tune.

After fitting new contacts, the gap should be re-set after about
500 miles, since the plastic heel of the moving contact assembly
will 'bed-in' slightly and close the gap.

5.4.2 **Contact lubrication**

Over-lubrication should be avoided. Any oil spreading across the
face of the contacts will carbonize and increase the electrical
resistance. This can be critical if an electronic ignition conversion
has been fitted, since the current in such cases is only a fraction of
an amp and a film of oil could cause a complete breakdown of
primary current flow. On the Lucas 25D4 all that is required is
one or two drops of oil at the following points:

(a) Through the hole (4) in the contact breaker plate to lubricate
the centrifugal weights.
(b) Around the screw (5) in the centre of the cam spindle. This
will percolate by gravity to the spindle bearings.

Carefully wipe away any surplus lubricant and check that the
contacts are clean and dry.

Before replacing the distributor cap the leads should be wiped
clean and the connections checked for tightness. The whole of the
distributor cap should be clean, but the most important area is the

inside surface, which should be wiped carefully with a clean dry cloth. The inside surface should then be examined for signs of cracking. If the cap appears to be in good order and there is no previous history of misfiring under load or at high speed it can be refitted. A simple test for a suspect distributor cap is give below:

(a) Detach all spark plug leads and the distributor end of the HT lead from the centre of the cap.
(b) Switch on the ignition.
(c) Rotate the engine until the points are closed.
(d) Insert the HT lead (from the coil) into each empty socket on the distributor cap in turn.
(e) At each position in turn give a quick flick to open the contacts, using an electrician's screw-driver (insulated handle), at the same time observing the inside surface of the distributor cap.

If tracking has been occurring a spark will be seen to jump across the surface of the cap. The reader will see that this test removes the normal plug gap from the HT circuit and subjects the insulation of the distributor cap to full HT voltage (circa 20 000 volts) in the reverse direction.

5.5 Operation no. 4: Ignition timing (6000 miles)

As a young apprentice, the author was shown how to time the ignition on the contemporary engines of the Twenties. A strip of cigarette paper was trapped between the contacts while a fitter's mate slowly turned the starting handle. The moment when the cigarette paper was released was taken to be the time of ignition. The actual crank-angle position could be checked by the marks on the flywheel. The reader will agree that this method was not very accurate, even though cigarette paper is very thin. A later method was to use a 'pea-lamp', i.e. a small 12 volt bulb, connected in series in the LT circuit. With the ignition switched on the engine was rotated slowly until the light went out, indicating the opening of the points. This is a perfectly accurate way of deter-mining the *static* timing. Even so the *dynamic* timing, the actual timing when the engine is running, is the true timing and this can differ from the static by 2–3 degrees even on a distributor with very little wear in the bearings.

5.5.1 **Dynamic timing**

The modern timing light costs less than £20 and is the only
accurate method available to the DIY tuner. Faced with the choice
of buying a contact-assisted electronic ignition system *or* a timing
light the author would choose the former, since many novice
tuners do not find the cheaper low-powered timing lights very
easy to use. The Boyer-Bransden electronic ignition system, for
example, includes a light-emitting diode in the circuit which
lights up when the contacts open. This gives a very reliable
method of measuring the static ignition timing. An accurate static
timing is always preferable to a doubtful dynamic timing.

The reader should therefore buy one of the more powerful
stroboscopic timing lights, one that gives good illumination of the
flywheel timing marks. Some of the more expensive ones can be
used in full sunlight. The modern timing light is very versatile
and can be used to observe the behaviour of the vacuum advance
capsule and the centrifugal advance mechanism.

The timing light should be connected across no. 1 plug lead and
a suitable earth. Some timing light makers supply a coiled spring
adaptor to fit to the plug or the distributor cap. Lacking this
adaptor it is not difficult to make one from a paper clip. One loop
of the paper clip is straightened and inserted in no. 1 terminal of
the distributor cap. With thin enough wire there is sufficient
space to pass between the cable and the plastic connector to make
contact with the conductor at the base.

The timing marks on the flywheel can usually be seen when the
inspection plate on the clutch housing has been removed. Sometimes a mirror is required to view the marks. On some engines
notches or grooves are made in the crankshaft pulley. These can
be checked against a stationary pointer. If in doubt one can
consult the Workshop Manual. Top Dead Centre on no. 1 and no.
4 cylinders is usually indicated by a 1/4 mark on the flywheel face,
preceeded by other marks, such as 10, 12, 16, etc to give other
degrees of advance. The Manual sometimes states the correct
timing as 'X degrees at 1000 rpm'. This engine speed must be
estimated, if the car is not fitted with a tachometer (1000 rpm is a
fast idle). Since vacuum advance sometimes begins at about this
speed it is advisable to disconnect the pipe to the vacuum capsule,

unless of course the Manual states otherwise.

If a darkened garage is not available and the flash from the timing light gives inadequate illumination of the marks on the flywheel one can try putting a patch of white paint on the flywheel in the appropriate area and, when the paint is dry, lining in the timing marks with a black pencil or felt pen.

When the existing timing has been ascertained, the distributor can be advanced or retarded until the recommended timing is given. To make small timing changes, the distributor clamp bolt should only be slackened until the distributor can be rotated by firm hand pressure. If the clamp is slackened too much the timing could change by drive reaction. When satisfied that the correct timing has been found the clamp bolt can be fully tightened. Some Lucas distributors are fitted with a knurled nut and a tangental screw at the base of the distributor body. Turning the screw by hand against the 'clicks' one can feel as the knurled nut is turned one can make very small changes in timing, i.e. approximately four or five clicks of the ratchet mechanism per crank-angle degree. One must thank the onset of emission control regulations for the disappearance of this simple device. It made ignition timing changes too easy!

5.5.2 Centrifugal advance

The majority of modern engines have a vacuum advance capsule mounted on the side of the distributor. They all carry a centrifugal advance mechanism. Before checking the behaviour of the centrifugal advance mechanism one should always disconnect the capillary tube on the side of the vacuum capsule, since we would be confused trying to interpret their combined action. With the vacuum capsule disconnected, the procedure is therefore as follows:

(a) Observe the initial timing mark at idle, using the strobe lamp. This of course has just been correctly set already.

(b) Increase the engine speed in graduated steps. If the throttle cannot be reached from the place where the strobe light is used an assistant is required to operate the accelerator. When the centrifugal advance mechanism is functioning correctly the timing light flashes will be seen to advance steadily from the fast idle timing mark and continue to advance up to some higher speed. The start of the advance will be at some speed above a fast idle and will continue in some cases up to a mid-range speed of, say, 2500 rpm. Some engines have distributors designed to give continuous advance up to higher speeds. The actual design advance characteristics are usually given in the Maker's Manual.

Excessive advance is usually caused by weakened balance weight springs. Insufficient, or jerky, advance movement indicates worn or sticking balance weights. A spot of oil dropped through the oil hole in the contact base plate sometimes cures a jerky advance movement. The majority of DIY tuners will probably need professional help to fit new springs or replace worn balance weights.

5.5.3 Vacuum advance

It is not easy to check the working of the vacuum advance in isolation, but the mechanism is simple and the usual failure is a punctured diaphragm. Check as follows:

(a) Re-connect the vacuum advance pipe.
(b) Observe the idle timing mark with the strobe light.
(c) Gradually increase the engine speed up to the point on the speed curve where centrifugal advance levels out. Since less advance should be given with reduction in manifold vacuum the combination of the two will be a negligible advance in comparison with the readings obtained in the centrifugal advance tests with the vacuum unit disconnected. A simple test for the operation of the vacuum capsule is a rapid opening of the throttle. The vacuum advance should operate immediately, then return to a more retarded position as the induction vacuum settles to a lower equilibrium value for the larger throttle opening. If no such timing movement is apparent the fault is almost invariably a leaking diaphragm in the capsule. Replacement capsules are not expensive and are easy to fit.

5.5.4 **Vacuum retard**

Some of the latest cars, designed to comply with the more stringent emission control regulations, are fitted with a second control capsule. This is designed to retard the ignition at very high manifold readings. This is necessary to reduce exhaust emissions during over-run (i.e. closed-throttle deceleration). It can be checked by setting down the idling speed to the lowest speed that will give steady running. When the pipe to this vacuum unit is disconnected the timing mark shown by the strobe light should advance by a few degrees.

5.5.5 **Diaphragm check**

A punctured diaphragm on either advance or retard capsule can be given a simple static test. The capillary tube should be disconnected. Then the contact base-plate should be rotated by hand against the action of the spring inside the capsule. Next the wetted finger is placed over the pipe union on the capsule to create a seal. The base-plate is released and, with a diaphragm in good condition the base-plate will be seen to rotate only part-way back. When the finger is released after three or four seconds the base-plate will return all the way to the starting position. If the diaphragm is punctured this second movement will not occur.

An ineffective vacuum advance capsule can have a disastrous effect on fuel economy.

5.6 **Operation no. 5: carburetter tuning (6000 miles)**

In the present context of maintaining an economical mixture this operation is confined to the adjustment of the idling mixture, a check on the correct working of such devices as the accelerator pump and any cruise economy devices and checking the level of the oil in the damper on variable venturi carburetters. These aspects of carburetter tuning are covered in the two succeeding chapters.

5.6.1 **The air cleaner**

It is normal to tune the carburetter(s) with the air cleaner removed, if only to make the carburetter adjusting screws more accessible. A dirty air cleaner element, however, always enriches the mixture and a very dirty air cleaner can give a very rich mixture and, simultaneously, a reduction in maximum power. Retention of a dirty air cleaner will therefore destroy all our efforts at economy tuning. Some older cars were fitted with air cleaner elements that could be cleaned and re-used. Modern engines, however, are geared to our throwaway society and are fitted with relatively cheap paper elements and it is essential to replace these at the recommended mileage. Retention for a longer mileage is a false economy.

5.7 **Operation no. 6: road testing (6000 miles)**

The Road Test, measured as Test no. 1 in the previous chapter, should be repeated. There will probably be a slight improvement in the acceleration figures, but a substantial gain in mpg can be anticipated. The values measured in this second test now become our Reference Road Test to be used for comparison in the future when we suspect a falling-off in economy.

6

Variable Venturi Carburetters

6.1 SU carburetters

The SU carburetter has been with us since the Skinner brothers formed the 'Skinner Union' or the SU Carburetter Company in 1910. The well-known Type H was with us unchanged for a generation. It could be said that the SU Company was a little self-satisfied at this time. The challenge of emission control regulations has been well met by the present generation of development engineers at SU and the latest SU carburetters are in the forefront of the latest economy drive.

These newest SU carburetters are very reliable and will hold their tune for at least 15 000 miles. As a general rule, however, we can say that the earlier the design the more frequently must it be tuned. With older cars, carburetters can eventually wear out, since carburetters with badly worn throttle spindle bearings will suffer air leaks that make accurate tuning impossible.

6.1.1 The type H

A good introduction to the SU variable venturi principle can be made through a study of the Type H. This design was still fitted to a few BMC models in the late Sixties. It has been made in horizontal and semi-downdraught form with sizes ranging from 1 1/8 inch to 2 inch bore. Fig. 6.1 is a cross-section of the Type H.

At idling speeds the lower piston (12) rests on the bridge (8). Two small protuberances – not shown – prevent complete closure

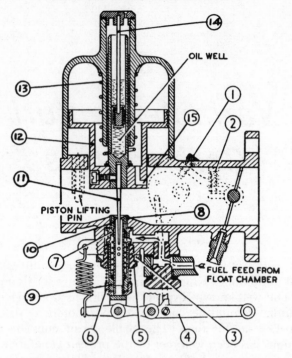

Fig. 6.1   Section of SU Type H carburetter.

in this position, thus giving a controlled air gap. On this model of
an SU carburetter, the air velocity through this slit was considered
to be high enough to give adequate pulverisation of the petrol
into a small droplet spray, even at idle air flows. Petrol enters this
narrow venturi through the annular space between the needle
(11) and the petrol jet (10). The height of this venturi piston and
the position of the needle in the jet is controlled at all air flows,
other than idle, by the forces acting on the larger diameter piston.
This piston is a free fit inside the suction chamber and is designed
not to contact the chamber wall, since the clearance here is greater
than that provided on the central guide. This lubricated guide
comprises a steel piston rod attached to the piston assembly,
which is a sliding fit on the central bore of the suction chamber.
    The suction chamber is maintained at a fairly constant de-

pression since it is balanced to the pre-throttle zone pressure by the drilled passage (15). The annular area below the large piston is open to atmospheric pressure. When the throttle-plate is opened and the depression in the zone between the venturi and the throttle tends to increase, this higher depression is immediately transmitted through passage (15) to the region above the suction piston. This causes the piston to rise until the effective venturi area has dropped the pre-throttle-zone depression to the original value. This value is by design a depression of 8 inches of water. This is about $\frac{1}{4}$ psi and therefore does not represent any great loss in ultimate power output. On modern carburetters, a light spring is provided to supplement the weight of the piston. The strength of this spring can be changed as a part of fine tuning procedure.

As the piston assembly rises and falls under the influence of the throttle position the tapered needle also rises and falls inside the petrol jet. By careful choice of the needle profile the mixture strength can be matched to the engine requirements. One could for example choose to give a constant air/fuel ratio at all times, or, in keeping with our knowledge of the real engine requirements, choose a rich mixture for idling, a lean mixture up to 80% of full power and a rich mixture again as we approach full power. The optimum dimensions of the needle are found by dynamometer testing at the maker's works.

Starting and cold running

To give a richer mixture for starting and cold running the Type H has a lever (4) operated by the manual choke control which lowers the jet relative to the needle. The jet (10) slides between an upper seal (7) and a lower seal (9). These seals are supposed to prevent leakage at both ends of the jet tube, but the SU Company would be the last to deny that they have sold many thousands of replacement seal kits over the years. If the base of the jet assembly, i.e. the adjusting nut (6) feels wet to the touch or is dripping, new seals are required. For starting, the jet lever (4) is pulled upwards, thus lowering the jet to give a very rich mixture. As the engine warms up this choke control is gradually pushed back to the normal running position.

Cruising economy

No special economy device is provided. SU exploit the natural phenomenon associated with the pressure fluctuations in the induction system to provide a lean part-throttle mixture. Since each induction stroke subjects the carburetter to an air flow that rises to a maximum as the piston passes its peak velocity and falls to a minimum as the inlet valve closes, the resultant flow is inevitably pulsating or wave-like. It can be shown that a pulsating flow across a petrol jet will induce a greater flow of petrol from the jet than a steady flow of identical total rate. With a part-closed throttle these pulsations are damped, i.e. the variations each side of the mean are much reduced. With a wide-open throttle the pulsations are undamped. Thus, *at a given air flow* and a wide-open throttle (say, when climbing a gradient), the mixture will be slightly richer than the same mean air flow with a part-closed throttle (say, when cruising at 50 mph).

Unfortunately, this phenomenon is more pronounced when fewer cylinders share the same carburetter. With one carburetter to six cylinders the mixture change from full throttle to part throttle is inadequate to give economical cruising and SU have in the past devised ingenious cruising economy devices to cover this special case.

Acceleration

When the throttle is opened suddenly to provide maximum acceleration it is essential that the mixture fed by the carburetter is enriched immediately to supply the additional fuel required, as discussed in Chapter 3. The device used on all models of SU carburetter is the piston damper. The damper works like a single-acting hydraulic suspension damper. The damper plunger is mounted on the end of a rod (14) and only gives a damping action when the suction piston is rising. A sleeve valve surrounds the plunger and is an easy fit in the steel guide bore. During acceleration when the piston assembly begins to rise, the oil drag between the bore and the sleeve carries the sleeve to its upper position where it forms a seal against a rubber O-ring. In the upward direction, the only passage for oil between the lower and upper parts of the oil well is the annular passage between the sleeve and the bore. In the opposite direction, the sleeve falls and exposes a free passage for oil on the inside of the sleeve. The overall effect of this damping action is to delay the rise of the piston assembly during acceleration. This delay creates a depression above the petrol jet that exceeds the normal 8 inches of water. The extent of this depression enhancement will vary with the speed of opening of the throttle, being negligible when opened slowly. The application of an enhanced depression to the jet gives an enchanced fuel flow, hence an enriched mixture.

Tuning the Type H

The SU carburetter only requires one little service between tune-ups. The oil has a tendency to disappear from the damper chamber. The level should be checked every 1000 miles.

Before tuning one should first remove the suction chamber, after first marking the correct position by pencil marks on the adjacent flanges. Despite the provision of an air cleaner it is still possible for carbonaceous matter to build up a light film on the lower venturi piston. This can usually be washed away with a wet petrol-soaked rag, before re-assembly.

When reassembled, free movement of the piston should be checked by raising the piston from below with one finger and allowing the piston to drop against the bridge. It should strike the bridge with an audible click. If the needle appears to be binding in the jet it can be re-centred as follows:

(a) Loosen the jet lock nut (5).
(b) Raise and drop the piston several times.
(c) When the piston appears to be dropping freely, i.e. making an audible click, carefully re-tighten the lock nut.
(d) Re-check that re-tightening has not disturbed the position by lifting and dropping again.

Occasionally, one cannot obtain free movement because the needle has been bent. In this case one can only obtain a replacement needle. They are not expensive and are almost impossible to straighten.

When finally satisfied with the free movement of the needle in the jet the damper reservoir should be refilled with SAE Grade 20 oil to the level indicated in Fig. 6.1. If overfilled, the surplus oil will eventually escape through the breather hole in the cap.

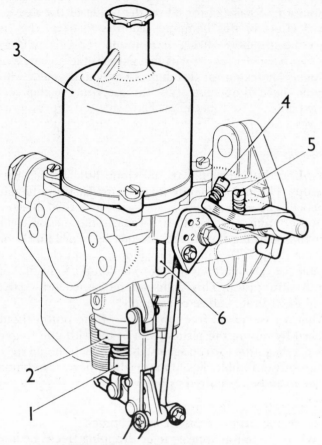

The Type H Carburetter

1. Jet adjusting nut.
2. Jet locking nut.
3. Piston/suction chamber.
4. Fast-idle adjusting screw.
5. Throttle adjusting screw.
6. Piston lifting pin.

Fig. 6.2 Positions of adjusting screws and jet adjusting nut on SU Type H carburetter.

The mixture adjustment of the Type H is made by screwing the jet adjusting nut upwards to weaken or downwards to enrich the mixture. The eight steps recommended by SU to tune the Type H are given in Figs. 6.3 to 6.10. A clear indication of the location of the adjusting nut and the other adjusting screws is given in Fig. 6.2.

The *correct* mixture setting is found as indicated in Fig. 6.8 when a raising of the lifting pin causes a slight increase in engine rpm followed by a return to the original speed. For *economy* with a small sacrifice in power the writer recommends that the adjusting nut should be turned up (anti-clockwise viewed from above) two or three flats.

Multi-carburetters

The tuning technique used with multi-carburetters is described in Figs. 6.11 to 6.15. For the perfectionist who is not too happy with the 'listening tube' there are several special balancing instruments on the market. An inexpensive device is 'The Jetsetter'. The address is given in the Appendix.

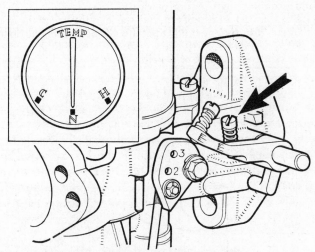

Tuning a Single Type H carburetter (Figs. 6.3–6.10)

Fig. 6.3
(a) Warm engine up to normal temperature.
(b) Switch off engine.
(c) Unscrew the throttle adjusting screw until it is just clear of its stop and the throttle is closed.
(d) Set throttle adjusting screw 1½ turns open.

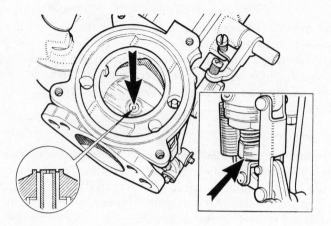

Fig. 6.4

(a) Mark for reassembly and remove piston/suction chamber unit.
(b) Disconnect mixture control wire.
(c) Screw the jet adjusting nut until the jet is flush with the bridge of the carburetter or fully up if this position cannot be obtained.

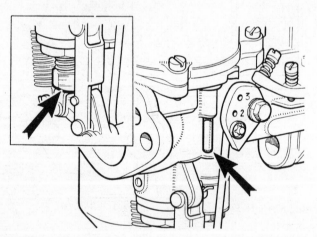

Fig. 6.5

(a) Replace the piston/suction chamber unit as marked.
(b) Check that piston falls freely onto the bridge when the lifting pin is released.
(c) Turn down the jet adjusting nut two complete turns.

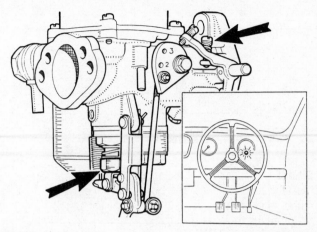

Fig. 6.6

(a) Restart the engine and adjust the throttle adjusting screw to give desired idling as idicated by the glow of the ignition warning light.

(b) Turn the jet adjusting nut up to weaken or down to richen until the fastest idling speed consistent with even running is obtained.

(c) Re-adjust the throttle adjusting screw to give correct idling if necessary.

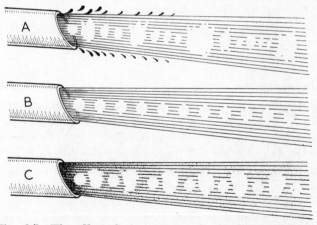

Fig. 6.7 The effect of mixture strength on exhaust smoke.

(a) Too weak: Irregular note, splashy misfire, and colourless.

(b) Correct: Regular and even note.

(c) Too rich: Regular or rhythmical misfire, blackish.

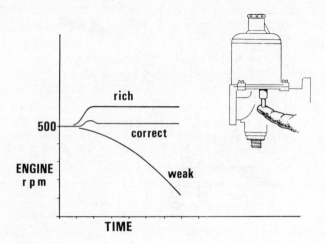

Fig. 6.8
(a) Check for correct mixture by gently pushing the lifting pin up about ¹/₃₂ in (0.8 mm).
(b) The graph illustrates the effect on engine rpm when the lifting pin raises the piston, indicating the mixture strength.

Rich mixture: rpm increase considerably.
Correct mixture: rpm increase very slightly.
Weak Mixture: rpm immediately decrease.

Another tuning method

The vacuum gauge technique is a well-tried method and from the author's experience it only fails on engines with extreme over-laps, such as one finds on sports/racing cars. The principle is based on the fact that an idling engine 'pulls' the highest vacuum against the nearly closed throttle when the mixture strength is adjusted to give maximum efficiency. With inefficient combus-tion, more air is consumed in supplying the power required at idle. When more air is consumed the throttle must be opened a little wider to maintain the same idling speed. This lowers the level of the induction vacuum. For a given idling speed, induction

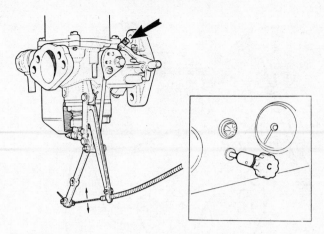

Fig. 6.9
(a) Reconnect the mixture control wire with about $^1/_{16}$ in (1.6 mm) free movement before it starts to pull on the jet lever.
(b) Pull the mixture control knob until the linkage is about to move the carburetter jet and adjust the fast-idle screw to give an engine speed of about 1000 rpm when hot.

vacuum is therefore highest at the most efficient mixture strength.

Many engines are already provided with a vacuum pipe fitting on the manifold for the attachment of the capillary tubing that actuates the ignition vacuum advance capsule. Since the vacuum advance is inoperative at idling speeds, the drilling in the carburetter flange being, by design, covered at this time by the throttle-plate edge, one can only use this point to measure the induction vacuum at a fast idle, i.e. at about 1000 rpm. When using this connection therefore one should insert a T-piece in the line. This will allow the vacuum advance capsule to operate correctly at a fast idle of 900–1000 rpm while reading the vacuum from the third branch of the T. Vacuum gauges are sold in accessory shops ('Command' by Narco National Ltd. and another by Speedograph Ltd.) and suitable adaptors and T-pieces are sold by

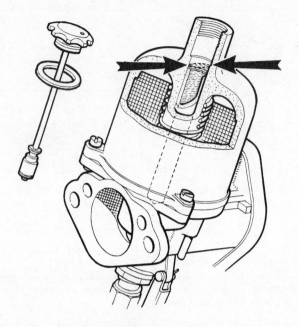

Fig. 6.10 Finally top up the piston damper with thin engine oil grade SAE 20 until the level is ½in (13 mm) above the top of the hollow piston rod.

N.B. On dust-proofed carburetters, identified by a transverse hole drilled in the neck of the suction chambers and no vent hole in the damper cap, the oil level should be ½in (13 mm) below the top of the hollow piston rod.

Speedograph Ltd. The latter firm also market a manifold flange adaptor that can be fitted between the carburetter flange and the manifold flange. This is an ideal arrangement since, when this has been fitted, one is free, to carry out the vacuum tuning method at the normal idling speed. The Command vacuum gauge is a dual purpose instrument since it also registers positive pressures and can be used to measure fuel pump pressures.

   The vacuum technique is extremely simple. When the vacuum

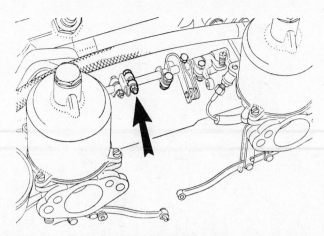

Fig. 6.11
(a) Remove the air cleaners and carry out the operation shown in Fig. 6.3 on all carburetters.
(b) Slacken one of the clamping bolts on the throttle spindle interconnections.
(c) Disconnect the jet control linkage by removing one or, in the case of triple carburetters, two of the linkage swivel pins.
(d) Carry out as in Figs. 6.4 and 6.5 for single carburetters, then additionally as in Fig. 6.12.

gauge has been connected, the thoroughly warmed engine is set to run at an idling speed of 600–700 rpm (or some slightly higher speed when the vacuum gauge is connected to the ignition vacuum advance tube). The jet adjusting nut is moved up and down, two flats of the hexagon at a time, until the point is found that gives the highest vacuum reading. On a typical touring engine, this will be found to be in the range 20–22 inches of mercury. Sports and racing engines with large valve overlaps usually have low combustion efficiencies at idle and give much lower values, often as low as 15 inches. On a normal touring engine a vacuum reading below 19 inches is an indication that some engine function, i.e. compression or ignition, is defective or badly tuned.

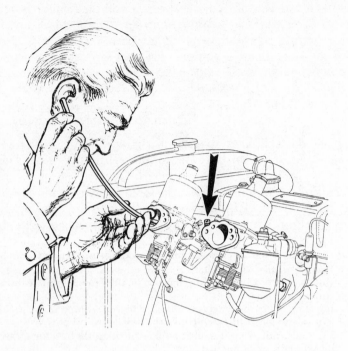

Fig. 6.12
(a) Restart the engine and adjust the throttle adjusting screws on
each carburetter to give the desired idling speed of 500 to 600 rpm
as recommended by the vehicle manufacturer.
(b) Compare the intensity of the intake 'hiss' on all carburetters
and alter the throttle adjusting screws until the 'hiss' is the same.

When the mixture setting that gives maximum vacuum reading
has been found we can try to stretch economy to the limit by
turning the adjusting nut two or three flats leaner (anti-clockwise
viewed from above). Experience in the final road test will show if
this leaner mixture is acceptable. If it results in occasional misfiring it cannot be economical.

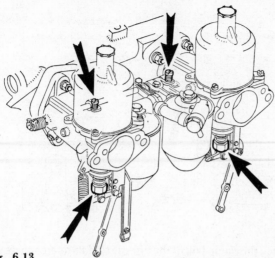

Fig. 6.13

(a) Turn the jet adjusting nuts on all carburetters up to weaken or down to richen the same amount until the faster idling speed consistent with even running is obtained.

(b) Re-adjust the throttle adjusting screws to give correct idling if necessary.

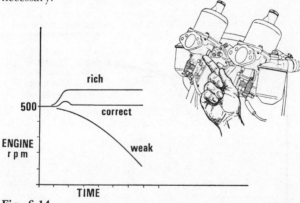

Fig. 6.14

(a) Check for correct mixture by gently pushing the lifting pin of the *front* carburetter up ¹/₃₂ in (0.8 mm.). The graph illustrates the possible effect on engine rpm.

(b) Repeat the operation on the *rear* carburetter and after adjustment re-check the front carburetter since the two are interdependent.

(c) Fig. 6.7 shows the correct type of exhaust smoke.

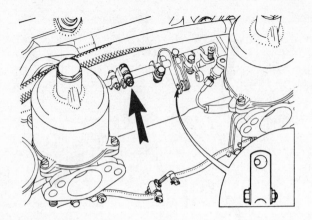

Fig. 6.15
(a) Tighten the clamp bolt of the throttle spindle interconnections and set the link pin lever with the pin resting against the edge of the pick-up lever hole (see inset). This provides the correct delay in opening the front carburetter throttle disc.
(b) Re-connect the jet control linkage, so that both jets commence to move simultaneously.

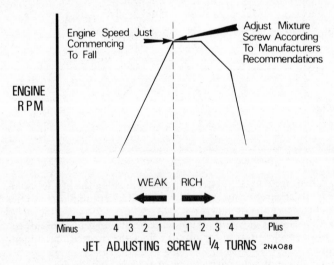

Fig. 6.16   Influence of mixture on engine rpm.

A third method

On an engine fitted with an engine rpm indicator this instrument can be used to tune the mixture. First the mixture setting is found which gives the highest reading on the rpm indicator for a given setting of the idle adjustment screw. The mixture strength is then leaned off step by step until the point is reached shown by the dotted line in Fig. 6.16. This correct economy setting is found when a small movement of the adjusting nut in the lean direction produces a small drop in rpm. Moving the nut in the opposite direction by the same number of flats should restore the maximum rpm. This is the most economical setting. If, at this stage, the idling speed is too high the idle adjusting screw should be re-set.

6.1.2 **The Type HS**

The Type HS was developed as an improved version of the near-vintage H Type. By connecting the float chamber to the main carburetter body by a length of plastic tubing the abominable leak-prone seals (7 and 9 in Fig. 6.1) disappeared. Another feature of the Type HS was the introduction of a spring-loaded needle, a perfect example of lateral thinking as advocated by Dr Edward De Bono. For years we had been urged by SU to prevent contact between the needle and the side of the jet. Now they designed one that incorporated a spring-loaded attachment to ensure that the needle was always pressed lightly against the side of the jet! On earlier designs an alignment mark was etched on the needle guide (see Fig. 6.17). Later models have a more positive V-groove in the needle guide. Obviously the old needle centring technique has no longer any significance. The origin of this spring-biased needle lies in some work carried out in the SU experimental laboratories when they were trying to reduce fuel metering variables under the challenge of the exhaust emmission regulations. It was shown that an incorrectly centred needle gave a slightly higher fuel flow than one where the needle was perfectly concentric in the jet (see Fig. 6.18). Since they could hardly demand the average motor mechanic to position the needle at the exact centre of the jet, the only alternative was to design a needle that was always in light contact with the side.

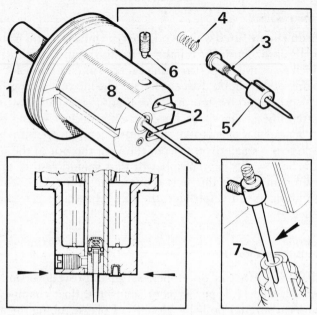

1. Piston rod	5. Needle guide
2. Transfer holes	6. Needle locking screw
3. Jet needle	7. Needle biased in jet
4. Needle spring	8. Etch mark

Fig. 6.17 The spring-loaded jet needle.

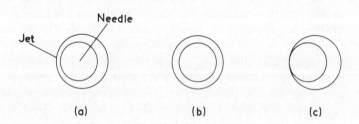

Fig. 6.18 Influence of needle centring on fuel flow. An off-centre needle (a) gives a higher flow than one perfectly centred (b). For consistent performance the latest carburetter has the needle spring-loaded to touch the jet wall (c).

Two other optional features appear on the Type HS. These are normally provided on cars intended for markets where emission control regulations exist. The first option is a temperature compensator, actuated by bi-metal discs, which raises and lowers the jet head to compensate for changes in fuel viscosity with temperature. The other is an over-run valve incorporated in the throttle-plate. This is a simple spring-loaded valve. It seldom gives any trouble and would not affect economy even if it did become stuck. Hydrocarbon levels tend to rise when the manifold depression rises above 24 inches of mercury. This could happen on the over-run and the purpose of the over-run valve is to admit an air bleed under these conditions to limit the rise in depression.

Tuning the Type HS

Any of the three tuning methods described above can be used on the Type HS. The levers, linkages and adjusting screws are in a more complex arrangement, but can be identified by reference to Fig. 6.19. On the HS2, HS4 and HS6 models the jet head can be raised and lowered to change the mixture strength as in the H Type. On the HS4C and HS8 a jet adjusting screw is provided at a higher level. This screw acts on a system of two levers plus a vertical link to move the jet head up and down. Clockwise rotation (viewed from above) enriches the mixture.

In some countries where vehicle emission standards are enforced it is mandatory to use an exhaust gas analyser to measure the CO levels in the exhaust when tuning the carburetter. In Europe, however, at the time of writing, there is no law that prevents an owner from tuning his own engine by any means at his disposal.

6.1.3 **The Type HIF**

This modern variant is usually fitted to medium and large capacity engines and incorporates additional features to give improved carburation at low speeds and a very tight control on mixture variations.

Concentric float chamber

So-called 'concentric' float chambers were first used on Weber

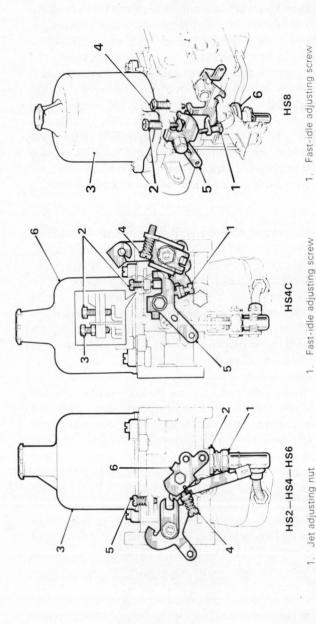

HS8

1. Fast-idle adjusting screw
2. Jet adjusting screw
3. Suction chamber assembly
4. Throttle adjusting screw
5. Cam lever
6. Temperature compensator

HS4C

1. Fast-idle adjusting screw
2. Throttle adjusting screw
3. Jet adjusting screw
4. Lost motion adjusting screw
5. Cam lever
6. Suction chamber assembly

HS2—HS4—HS6

1. Jet adjusting nut
2. Jet locking nut
3. Suction chamber assembly
4. Fast-idle adjusting screw
5. Throttle adjusting screw
6. Cam lever

Fig. 6.19 Position of adjusting screws on SU Type HS.

carburetters. In the Type HIF a horse shoe-shaped float is used. Since the jet assembly lies at the 'centre' of this horse shoe, variations in hydrostatic head on the jet from fore and aft or side to side cornering accelerations are minimized. In the past, such refinements were only considered necessary on racing cars.

Part-throttle by-pass emulsion

Although the variable venturi concept is acknowledged to give much better fuel atomization at low throttle openings than that afforded by fixed venturi carburetters it was still considered advisable in this latest design to incorporate a part-throttle by-pass system into the Type HIF carburetter to improve the quality of fuel atomization at low engine speeds. A small-bore passage carries the mixture created at the top of the jet to a discharge point at the throttle-edge. This small-bore passage maintains a very high gas flow to prevent fall-out or the agglomeration of large droplets.

Cold start enrichment

The third change from the older H and HS designs is the provision of an augmented fuel supply controlled by a rotary valve as a means of cold start enrichment.

Tuning the Type HIF

Any of the three tuning methods recommended for the Type H can be used on the Type HIF. In Fig. 6.20 the throttle adjusting screw (4) and the fast-idle adjusting screw (5) are indicated. The jet adjusting screw (2) is sunk inside a boss, as shown. On emission control carburetters a metal cap covers this screw, but this is not difficult to extract. It is possible that the local agent has already removed this cap since not all engines have been correctly tuned when they leave the factory. Clockwise movement of the jet adjusting screw will give mixture enrichment.

6.2 Further economy tuning

The latest SU carburetters are very accurate metering devices and there is little we can do, apart from careful tuning, to extract

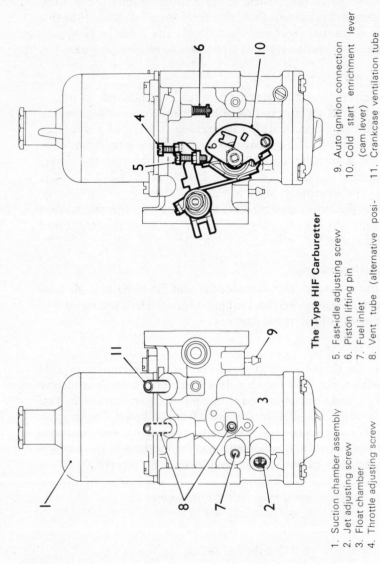

The Type HIF Carburetter

1. Suction chamber assembly
2. Jet adjusting screw
3. Float chamber
4. Throttle adjusting screw
5. Fast-idle adjusting screw
6. Piston lifting pin
7. Fuel inlet
8. Vent tube (alternative positions)
9. Auto ignition connection
10. Cold start enrichment lever (cam lever)
11. Crankcase ventilation tube

Fig. 6.20 Position of adjusting screws on SU Type HIF.

that last drop of economy. Earlier models, particularly those made before the introduction of the spring-loaded needle, were given a more flexible specification. Sometimes a choice of needles is specified, rich, standard and weak. Two piston springs are sometimes specified. These variations were provided at the time to give some flexibility to individual tuning. Some mass-produced engines breathe better than others. Variations can come from such features as a slight mismatch where induction manifold ports meet the ports in the cylinder head or from small variations in other dimensions. One would not anticipate such variations in an engine made by Rolls-Royce, but we note that the Phantom V and VI are given a choice of two piston springs. The use of the blue spring will give a slightly weaker mixture throughout the whole operating range than the red one.

The recommended needles and springs for the whole range of makes and models of cars fitted with SU carburetters can be obtained from SU Carburetters, Wood Lane, Erdington, Birmingham. The publication is AUC 9631 M.

Another booklet of value to the keen engine tuner is the complete list of needle sizes (publication AUC 9618). This second publication, however, is usually only of value when an engine has been modified by a previous owner; perhaps fitted with a special cylinder head with larger valves or with a manifold to take larger carburetters. Such major changes usually call for a change in needle profile. If any tendency to over-richness is suspected in part of the operating range (the previous owner having only been interested in an increase in maximum power) one can experiment with alternative needles chosen from the extremely large selection given in publication AUC 9618. The local SU stockist can sometimes supply these publications.

6.2.1 Needles

When tuning a perfectly standard engine one can always buy a weak needle, if one is listed, and give it a trial. Road tests will soon show if the mixture is too weak. Listen for any tendency to pink under load and examine the plug insulator noses for the white condition that indicates overheating.

6.2.2. **Springs**

Piston springs are identified by a colour code painted on the end coils of the spring. The range covering carburetter sizes up to $1\frac{3}{4}$ inch bore is: blue $2\frac{1}{2}$ oz, red: $4\frac{1}{2}$ oz, yellow: 8 oz, green: 12 oz.

When testing the effect of a spring change one should remember that the lighter spring always gives a leaner mixture (the opposite effect to what one might expect).

6.3 **Zenith-Stromberg carburetters**

6.3.1 **The Type CD**

The Stromberg Constant Depression (CD) carburetter is made in Great Britain by the Zenith Carburetter Company Ltd. Early examples carried the name Stromberg, but more recent models are labelled Zenith.

The CD carburetter operates on the same principle as the SU. It does in fact revert to the original concept of the carburetter patented in 1905 by George H. Skinner which used a leather bellows to seal the constant depression chamber. The leather bellows was not reliable and Skinner would have been amazed at the durability of the modern petrol resistant elastomers from which the Type CD diaphragms can be made. Fig. 6.21 is a cross-section of the Type CD carburetter. The piston in the SU carburetter is replaced by a synthetic rubber and fabric diaphragm (L) which is sealed on the outside by the flange of the diaphragm chamber (M) and is attached by screws to the top of the air-valve or piston (K). The diaphragm chamber is connected by the transfer hole (N) to the constant depression zone between the air-valve and the throttle-plate.

As in the SU carburetter the air-valve or piston attains an equilibrium level above the bridge as dictated by the balance of forces acting on the air-valve and diaphragm assembly. Downward forces are the weight of the assembly plus the force from the return spring (P). Upward forces to create the balance are supplied by the depression in the suction chamber multiplied by the effective diaphragm area. The damper is identical to that used by SU, being designed to resist upward movement of the

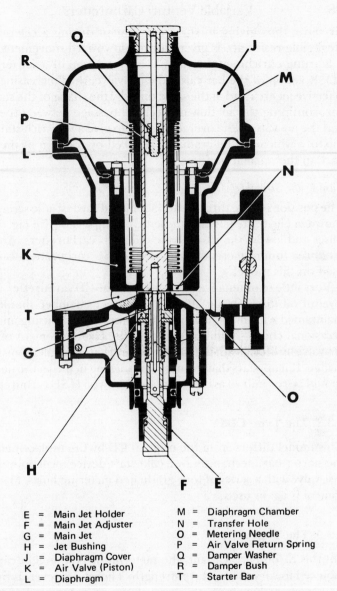

E = Main Jet Holder	M = Diaphragm Chamber
F = Main Jet Adjuster	N = Transfer Hole
G = Main Jet	O = Metering Needle
H = Jet Bushing	P = Air Valve Return Spring
J = Diaphragm Cover	Q = Damper Washer
K = Air Valve (Piston)	R = Damper Bush
L = Diaphragm	T = Starter Bar

Fig. 6.21 Cross-section of Zenith-Stromberg Type CD.

air-valve, thus giving mixture enrichment during acceleration. Negligible resistance is given during downward movement.

Starting enrichment on the CD is by rotation of a starter bar (T). Rotation of the bar raises the air-valve, thus increasing the effective jet area and at the same time, by the shape of the starter bar, confining the air flow to a central passage between the bar and the air-valve. The overall effect is to give a very rich starting mixture which can be gradually reduced by rotation of the bar back to the normal position.

Tuning the Stromberg CD

The positions of the throttle-stop screw and the fast-idle screw are shown in Fig. 6.22. Turning the jet adjusting screw (F in Fig. 6.21) raises and lowers the jet as in the older SU carburetters. Any of the three tuning methods proposed for SU carburetters can be used on this model.

Very little maintenance is required on the CD carburetter. The level of oil (SAE Grade 20) in the damper chamber should be maintained within about ¼ inch of the top of the housing and an occasional check should be made on the free movement of the air-valve by lifting with the forefinger and letting it fall against the bridge. If it appears that the needle is sticking in the jet the needle should be re-centred as described for the Type H SU carburetter.

6.3.2 **The Type CDS**

This model differs from the original CD by the replacement of the starter bar mechanism by a cold-start device incorporating a disc valve with a series of four graduated metering holes. Manual control is again used.

6.3.3 **The Type CD3**

On this Stromberg variant, the metering needle height can be adjusted to vary the mixture strength. The jet, however, is fixed. As shown in Fig. 6.23, the base of the oil reservoir has been removed, oil seepage into the metering zone being prevented by an O-ring seal. The needle is spring-loaded, as in the Type HS SU carburetter. Rotation of the spring carrier is prevented by the

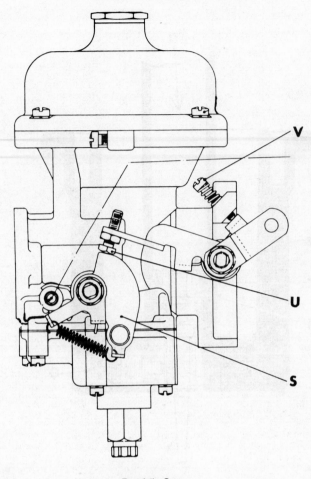

S = Fast-Idle Cam
U = Fast-Idle Screw
V = Throttle Stop Screw

Fig. 6.22 Position of adjusting screws on Zenith-Stromberg Type CD.

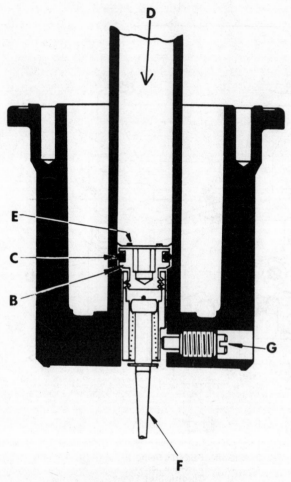

B = Needle Adjusting Screw
C = 'O' ring for Adjusting Screw
D = Oil Reservoir
E = Retaining Washer
F = Metering Needle
G = Spring-Loaded Locating Screw

Fig. 6.23 The adjustable needle in the Zenith-Stromberg Type CD3.

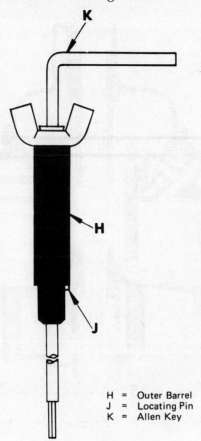

Fig. 6.24   The mixture adjusting tool for the Zenith-Stromberg
Type CD3.

locating screw G. A special adjusting tool (see Fig. 6.24) may be
purchased from the local Zenith-Stromberg stockist. A long Allen
key (K) is passed through the outer barrel (H) to engage a socket
in the head of the adjusting screw. The pin (J) at the base of the
outer barrel locates in a slot to prevent the air-valve from turning
when the Allen key is rotated. Clockwise rotation gives a leaner
mixture.

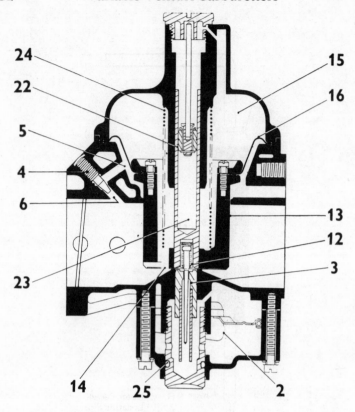

Fig. 6.25 Cross-section of Zenith-Stromberg Type CDSE.

6.3.4 The Type CDSE

This range of carburetters was developed from the Type CDS for use in countries where the strictest anti-pollution regulations are in force. In Europe, our regulations are not as severe as, for example the U.S.A. or Japan, but we can benefit from the excellent development work carried out by Zenith to improve the metering accuracy of their carburetters. When we find that our particular vehicle is fitted with such a carburetter it is *perfectly legal* for us to tune the vehicle to give an economical mixture to the best of our ability.

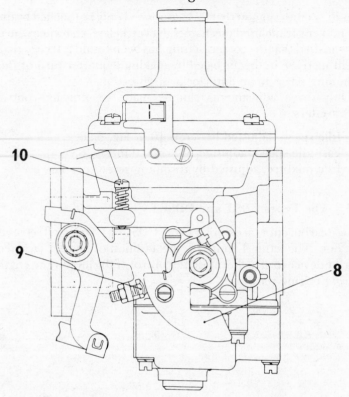

Fig. 6.26 Position of adjusting screws on Zenith-Stromberg Type CDSE.

Tuning the CDSE

In general, on a car which has done less than 20 000 miles from new, one seldom finds that the Type CSDE carburetter needs any attention, beyond the customary topping up of the damper chamber. If one has any doubts that the initial tuning was conscientiously carried out, one should insist that the supplier should correct the negligence.

Zenith provide an 'idle trimming screw' (4 in Fig. 6.25). In Zenith's own words this is 'a very fine adjustment to compensate for differences between a new "tight" engine and one that is run-in'. Zenith suggest that this screw need only be touched by the car maker's franchised mechanics. Nevertheless, an owner who is not satisfied that the correct setting has been found is free to try a slight increase in the air bleed by making a quarter turn of this trimming screw in an anti-clockwise direction.

Only three adjustments can be made to emission-control carburetters:

(a) Idle speed, adjusted by screw 10 in Fig. 6.26.
(b) Fast-idle speed, adjusted by screw 9 in Fig. 6.26.
(c) Idle mixture, adjusted by trimming screw 4 in Fig. 6.25.

6.3.5 The Types CDST and CDSET

These carburetters are similar to the CDS and CDSE is all respects but one. The letter T indicates the addition of a water-jacketed starter device to give thermostatic control of the mixture strength during engine warm-up.

7

Fixed Venturi Carburetters

The same general rule applies to most fixed venturi carburetters as was observed in the variable venturi designs. The more recent the design, the less does tuning become necessary or desirable. Older carburetters were not always well protected against the ingress of dirt. Float bowls can collect sediment and jets can occasionally choke, especially the smaller pilot jet. Moreover, with age, one must not be surprised to find that rubber diaphragms fitted to accelerator pumps or to economy devices have begun to leak. These older ones are therefore described first. We must in all fairness give a word of warning. Carburetters are sometimes very complex mechanisms. Accelerator pumps have tiny steel balls that act as non-return valves, economy devices often have fiddly little springs. Even experienced mechanics sometimes drop these on the garage floor and spend ages trying to recover them. If you have doubts regarding your ability to dismantle a carburetter – don't!

There are usually three simple adjustments anyone who does not possess a natural talent for disaster can make:

(a) The idle speed.
(b) The fast-idle speed.
(c) The idle mixture strength.

7.1 Zenith carburetters

Some older cars are fitted with the original fixed venturi Zenith,

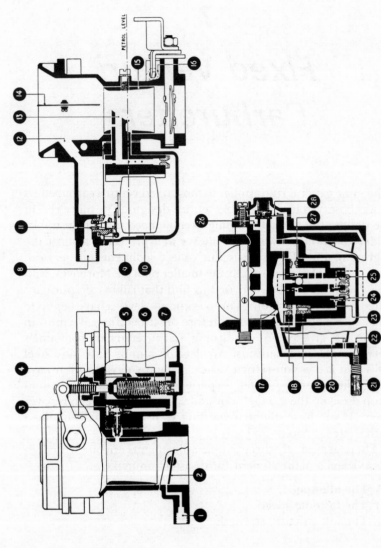

Fig. 7.1 Three cross-sections of Zenith Type 34VN.

for which the CD (Stromberg) design and its more modern developments are a more efficient replacement. As a typical example of the Zenith fixed venturi carburetter we have chosen the Series VN. There have been many other Series, since Zenith were making fixed venturi carburetters before many of us were born. Obviously, we cannot describe them all in this book.

7.1.1. The Series VN

The Zenith system of compensation was described in Chapter 3. In older designs of Zenith, the capacity well is open to the atmosphere. In the VN and subsequent designs, the admission of air to the well is controlled by one or more air bleeds. In the 34 VN shown in three cross-sections in Fig. 7.1, the main air bleed (27) is described as 'the ventilation screw to the capacity well.' Between this air bleed and the carburetter inlet zone is situated a second restrictor under the control of the economy device. Under cruising conditions, with a fairly high vacuum in the induction manifold, the diaphragm (28) is pulled to the right, thus giving an unrestricted passage to admit air to the ventilation screw (27). At full throttle, with a low depression acting on the area to the right of the diaphragm, the spring will push the diaphragm to the left, thus sealing off the large-bore passage. The air passage in series with the ventilation screw now becomes a small air bleed (26). The depression in the capacity well is therefore greater at full throttle than under part-throttle conditions. This gives a richer mixture to protect the valves and plugs when the engine is driven hard, but allows the use of a lean economical mixture when cruising.

A simple piston-type acceleration pump is used with non-return ball valves at inlet and exit. There are two positions for the cast block (4). In Fig 7.1 the block is shown in the long-stroke or Winter position. If we turn the block through 180 degrees we reach the Summer position. This setting reduces the amount of fuel injected during acceleration and must therefore help to reduce overall fuel consumption. Not all Zenith carburetters incorporate a means of reducing the amount of fuel injected by the acceleration pump, but the reader is well advised to check on this possibility.

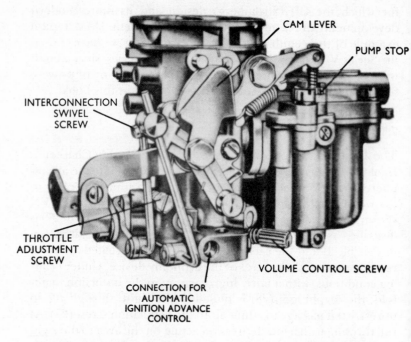

Fig. 7.2   Position of adjusting screws on Zenith Type 34VN.

Tuning the Series VN

Tuning is normally confined to adjusting the idling mixture by means of the volume control screw (21 in Fig. 7.1). For clarity, an external view of the carburetter is given in Fig. 7.2. The vacuum gauge technique described in Chapter 6 can be used to find the correct setting of the idling mixture. Alternatively, one can use the third method if an engine speed indicator is fitted.

On the more recent type of Zenith carburetter, the idle mixture adjustment screw controls the supply of a very rich petrol/air mixture that is fed downstream of the throttle plate to enter a stream of air. Thus, a leaner mixture is given when the screw is turned clockwise. On some earlier types the adjustment screw controls the amount of air only. Clockwise rotation in this case admits less air and enriches the idling mixture. Identification of the two types is easy on all types of Zenith downdraught carburetter. If the adjusting screw is nearer the top of the carburetter body than the bottom, it is an air regulation screw. If it is positioned near the bottom flange it is a volume control screw, i.e. a rich mixture screw as fitted to the Series VN.

7.2 Solex carburetters

Since Solex and Zenith carburetters and those originally called Stromberg are all sold in Great Britain by the Zenith Carburetter Company one might expect some measure of rationalization to appear with a reduction in the range of models produced. The CD range has now replaced the older fixed venturi Zenith, but the Solex range of fixed venturi carburetters, made largely in France and Germany, is still extensive.

7.2.1 The Type 32 BI

The well-established Solex layout can be seen in the cross-section

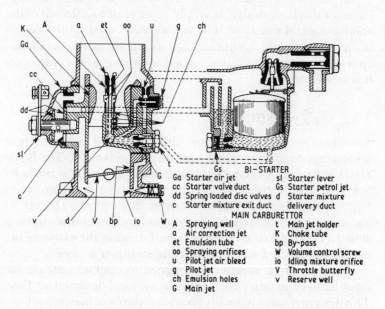

BI–STARTER	
Ga Starter air jet	sl Starter lever
cc Starter valve duct	Gs Starter petrol jet
dd Spring loaded disc valves	d Starter mixture
c Starter mixture exit duct	delivery duct

MAIN CARBURETTOR	
A Spraying well	t Main jet holder
a Air correction jet	K Choke tube
et Emulsion tube	bp By-pass
oo Spraying orifices	W Volume control screw
u Pilot jet air bleed	io Idling mixture orifice
g Pilot jet	V Throttle butterfly
ch Emulsion holes	v Reserve well
G Main jet	

Fig. 7.3 Cross-section of Solex Type 32BI.

in Fig. 7.3. As described in Chapter 3, the fuel/air mixture is supplied to the spraying orifices, too, through the main jet, G and the mixture strength compensation is controlled by the air correction jet, a. Fig. 7.3 is not a true cross-section, dotted lines being used to indicate the internal passages that carry fuel from the float chamber to the main jet, G, and to the Bi-starter carburetter on the opposite side. The emulsion tube, et, is held in place by the air correction jet, a. The pilot jet, g, draws fuel from the reserve well, v, and the idling mixture is composed of fuel metered by the pilot jet and the pilot jet air bleed, u. This very rich mixture passes down the vertical duct to the idling mixture orifice, io. The quantity of idling mixture entering the main air stream is controlled by the volume control screw W.

As the name implies, the Bi-starter is a two position starter carburetter with the choice of two enrichment levels for starting from cold. Over the years, this primitive two-position rotary valve has been developed into a multi-hole valve and finally into a thermostatically controlled automatic starter carburetter.

Normal tuning of the BI Series, and the subsequent developments of this basic design, is simply a matter of adjustment of the volume control screw, W. The two methods used on the Zenith carburetter, i.e. the vacuum gauge technique or the maximum rpm technique, should give a good approach to the most economical setting.

7.2.2 The Type SDID

Some continental cars are fitted with compound carburetters in an attempt to achieve good part-throttle economy. The Type SDID is typical of these compound instruments. It is made by Solex France. A simple 'butterfly' or plate type strangler, V1 in Fig. 7.4, is used for starting enrichment on the primary barrel. At low engine speeds the left-hand secondary barrel is completely closed. A spring-loaded poppet valve, C1, limits the extent of the depression created when the starting strangler is closed.

This type of carburetter is designed to operate only on the right-hand or primary venturi below a critical designed air flow. This designed value is usually about two-thirds of maximum flow. The primary throttle is connected directly to the accelerator linkages, but the secondary throttle only opens when the depression acting below it is sufficient to overcome a spring-loaded control. There are other ways of controlling the opening of the secondary circuit. In the PAIA Series, for example, a vacuum-operated diaphragm is used to open the secondary throttle. The vacuum pick-up point is in the waist of the primary venturi. The extent of this vacuum is therefore a direct measure of the air flow through the primary venturi.

Only one volume control screw (W in Fig. 7.4) is usually provided on compound carburetters.

7.2.3 The Type PHH

Where power, rather than economy, is the aim of the engine designer many of the more sporting cars are fitted with twin-barrel carburetters. Effectively, these are twin carburetters sharing the same float chamber. A typical example manufactured

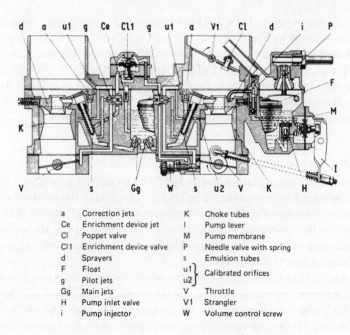

a	Correction jets	K	Choke tubes
Ce	Enrichment device jet	I	Pump lever
Cl	Poppet valve	M	Pump membrane
Cl1	Enrichment device valve	P	Needle valve with spring
d	Sprayers	s	Emulsion tubes
F	Float	u1	Calibrated orifices
g	Pilot jets	u2	
Gg	Main jets	V	Throttle
H	Pump inlet valve	V1	Strangler
i	Pump injector	W	Volume control screw

Fig. 7.4 Cross-section of Solex Type SDID.

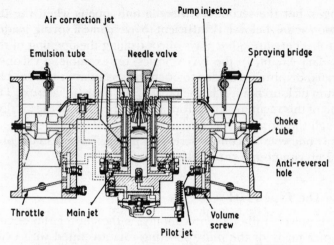

Fig. 7.5   Cross-section of Solex Type PHH.

in Germany is the Type PHH, a horizontal Solex design seen in cross-section in Fig. 7.5.

On this type of carburetter, there is a separate volume control screw for each venturi. Before attempting to tune the idling mixture by means of these two screws it is essential to balance the air flows passing the two throttle-plates at idle in the manner described for SU carburetters in the previous chapter.

7.3   Weber carburetters

The Weber has a glamorous image from its early association with the world of motor racing. The DCOE is typical of the established Weber horizontal twin-barrel carburetter that was fitted to many sports and racing cars in the past. More recent applications are the Lotus twin-cam engine used in Ford Escort and Cortinas and in the Lotus S4 sports car.

All Webers use the same system of compensation as the Solex, the tendency to enrichment from the main jet with increasing air flow being corrected by an air correction jet and an emulsion tube. All Webers also use a small diameter auxiliary venturi into which the mixture is discharged before admixing with the outer band of air passing between the main venturi and the auxiliary. This use of an auxiliary venturi gives improved atomization.

7.3.1 The Type DCOE

Access to the main jet (5 in Fig. 7.6), the emulsion tube (12), and the air correction jet (11) is obtained by removal of a circular cover held in place by a central wing-nut. The idling mixture is enriched by anti-clockwise rotation of the spring-loaded screw behind the carburetter mounting flange.

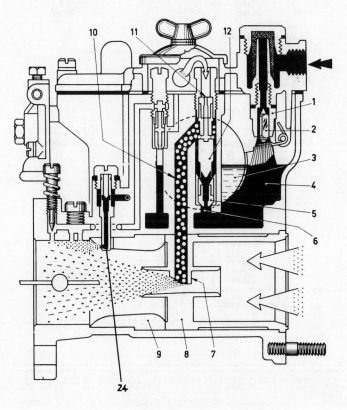

Fig. 7.6 Cross-section of Weber Type DCOE.

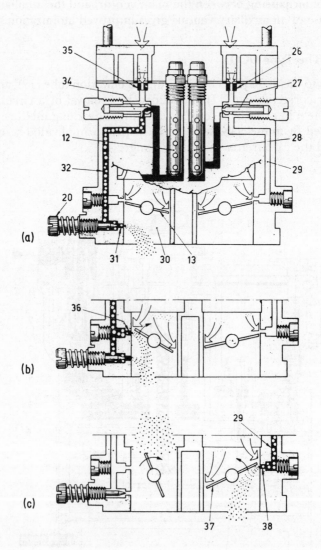

Fig. 7.7 Cross-section of Weber Type DCD, showing progression from primary venturi only to the opening of the secondary venturi.

7.3.2 **The Type DCD**

Weber make a range of compound carburetters. The DCD as fitted to the Cortina 1500GT and the Capri is typical. As in the Solex compound design (see Fig. 7.7), the idling mixture can only be adjusted on the primary barrel by a simple volume screw (20). As the throttle is opened further the progression mixture holes (36) are uncovered (only one hole shown in the illustration). Opening of the secondary throttle is entirely automatic, opening when the depression in the primary venturi throat reaches a predetermined value as shown in the lower illustration. To prevent any hesitation from an over-lean mixture at the critical point when the secondary throttle begins to open, fixed progression holes (38) are provided at the secondary throttle-edge. No adjustment is provided at this point. In rare cases where a tendency to hunt or hesitate occurs when the secondary throttle is opening, it is possible to effect a cure by an increase in the size of the secondary slow run jet (27). In the majority of cases any tendency to hesitate, where this has not been experienced in the past, is almost sure to be caused by a malfunction in the ignition system. If all else fails, the cleanliness of all the carburetter jets should be checked.

8

Proprietary Tuning Kits

8.1 **Colorplugs**

Gunson's Colorplugs are well known, but their other instruments, of more recent manufacture, can be combined with the Colorplug to produce a complete tune-up kit. The Colorplug, however, is so long established that we consider it worthy of a section to itself, even though it only gives a method of carburetter tuning.

8.1.1 **The Colortune 500 Kit**

The original Colorplug comprised a spark plug with a cylindrical transparent chamber above it through which the colour of the combustion flame could be viewed by the observer. A yellow flame indicated a rich mixture, an intense whitish-blue one a lean mixture. In the latest Colortune 500 Kit the colour of the combustion process is observed down a long tube, the Viewerscope. An angled mirror is provided for those cases where plug location prevents direct viewing.

The technique is straightforward. As planned in Chapter 5, carburetter tuning is the final operation. When the engine is warmed to normal running temperature the car should be parked in the shade since the colour of the flame will not show up well in a strong light. If the car is garaged to achieve this, care should be taken to allow the free escape of the poisonous exhaust fumes. The engine should be switched off and the Colortune plug fitted in place of one of the spark plugs. The adaptor lead and Viewerscope is then fitted, following the maker's instructions.

The procedure is as follows:

(a) With the engine running at the normal idling speed (about 700 rpm) the idle mixture screw is turned until a yellow flame can be seen through the Viewerscope. This indicates a rich mixture.

(b) The mixture screw is then turned in the opposite direction until a bunsen blue colour first appears. This is close to the maximum power mixture strength.

(c) By leaning the mixture still further one can find a setting where the colour is more of a whitish blue (see Fig. 8.1). This is the maximum economy setting.

(d) If the idling speed has changed this should be re-adjusted to about 700 rpm and operation (c) repeated.

The Colortune 500 plug and Viewerscope are made from special materials to withstand high temperatures. On high performance engines, however, care should be taken not to run the engine at speeds above 1000 rpm. For all kinds of engine, operations at speeds above normal idling speed should be limited to two minutes.

8.1.2 Multiple carburetters

Before using the Colorplug, idling air flows should be balanced by one of the methods described in Chapter 6. When this has been accomplished the Colorplug should be used in turn on one of the appropriate cylinders. For example, on a 4-cylinder twin-carburetter engine the Colorplug would first be fitted to no. 1 or 2 cylinder while the front carburetter was tuned, then fitted to no. 4 or 3 cylinder during the tuning of the rear carburetter. A triple carburetter installation would, correspondingly, require three Colorplug readings.

8.2 The full Gunson tune

Four products are recommended by Gunson's Colorplugs Ltd. as

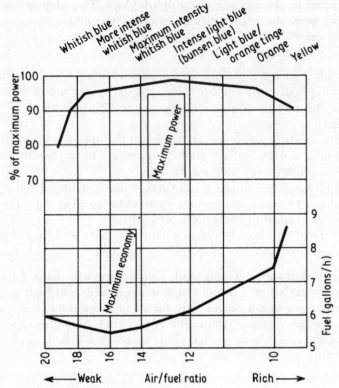

Fig. 8.1 Combustion flame colour as a guide to mixture strength.

a complete tuning kit. These four instruments are: Sparktune 2; Tachostrobe; Carbalancer; and Colortune 500. The latter has already been described.

8.2.1 Sparktune 2

The measurement of dwell angle as a scientific method of setting the distributor contacts was pioneered in Great Britain by Messrs Crypton. Many types of dwell meter are now made for the DIY tuner. Sparktune 2, the Gunson dwell meter also serves as a voltmeter. Dwell angle is the angle of distributor cam rotation when the contacts are *closed*. With inadequate dwell angle at high engine speeds there is insufficient time for the build-up of a high voltage in the coil. The correct setting of the contact gap is one approach to achieving this correct dwell angle, since the correct gap setting is designed to give the correct dwell angle. The dwell meter, however, has been shown to give greater accuracy.

The Sparktune 2 instrument is first connected as shown in Fig. 8.2. With a positive earth system the red and black leads shown in Fig. 8.2 should be reversed. The ignition is switched on and the engine turned over by the starter (no danger of starting the engine since the CB lead has been disconnected). While the engine is turning the instrument is adjusted to read 'S' by finger-tip movement of the control wheel, K.

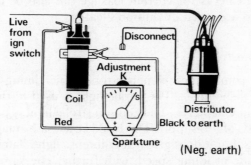

Fig. 8.2 Connections to *Sparktune 2* when setting instrument zero.

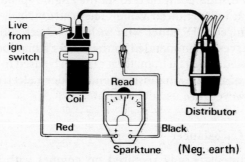

Fig. 8.3 Connections to *Sparktune 2* when measuring dwell angle.

The instrument is now 'zeroed' and can be connected as shown in Fig. 8.3 (remembering to reverse the connections when the electrical system has a positive earth).

We can now read the dwell angle. The engine is turned over on the starter while a reading is taken on the dwell meter. The correct reading for the particular make and model is given in a chart supplied with the instrument. If the reading is too high the contact gap should be increased; if too low it should be decreased.

The voltmeter component of Sparktune 2 can also be used to check the voltage drop across the contacts when fully closed. Dirty or pitted contacts are indicated by a voltage drop exceeding 0.2 volts. This voltmeter can also be used to reveal excessive resistance in other parts of the electrical circuits, i.e. in the starter circuit, the battery or any of the electrical components in the rest of the car. Checks for current leakage can also be made. Full instructions are given in Gunson's leaflet.

8.2.2 Tachostrobe

This is a combination xenon stroboscopic timing light and an optical tachometer. The strobe light is used in the manner described in Chapter 5. The Gunson strobe light is powered by 220–240 AC Mains supply. This could be a disadvantage to some DIY tuners, but it does give a very intense light. Timing checks can be made at idling speed, with further checks on vacuum advance and centrifugal advance at higher speeds.

The optical tachometer uses a moving shadow technique (described in Gunson's pamphlet) to set the engine speed extremely accurately at any required value. The instrument is invaluable to the advanced DIY tuner who wishes to check the automatic advance curves recommended in the maker's handbook. When the car is fitted with a tachometer it can also be used to carry out a calibration of the existing instrument. Modern electronic tachometers, however, are usually accurate.

8.2.3 The Carbalancer

This instrument is only required on engines with more than one carburetter. As with all such devices the technique involves placing a fixed orifice at the intake passage of each carburetter in turn and adjusting the idling adjusting screws to give identical readings on a flowmeter. To cope with a range of engine sizes the Carbalancer is provided with a 'micro-metering head'. This allows the size of the orifice to be adjusted to the required air flow for a fast idle speed. The same orifice setting must be used for all carburetters. The Carbalancer can be adjusted to fit most carburetters (side-draught or down-draught) with air intakes from 1 inch to $2\frac{1}{4}$ inch bore.

8.2.4 Colortune 500

This fourth step in the Full Gunson's Tune has already been described and can, of course, be used as an adjunct to any other tuning procedure. In any system of tuning, however, it is advisable to leave the carburetter tuning till last. It should never precede gap setting or ignition timing.

8.3 Command tune-up kits

Narco National Ltd market a 3-piece and a 4-piece tune-up kit, each supplied with detailed instructions. A special tune-up guide to assist trouble-shooting is also available. Command accessories are stocked by leading accessory shops. The 3-piece kit comprises a combined vacuum gauge and fuel pump tester, a compression tester and a Neon timing light. The timing light incorporates a remote starter switch. The fourth item in the 4-piece kit is a separate remote starter switch. By making this independent of the timing light it can be used when carrying out other operations, such as adjusting the valve clearances and measuring compression pressures.

Command also make a wide range of more sophisticated diagnostic analysers, dwell/tach meters and a very reasonably priced exhaust gas analyser.

9
Electronic Ignition Kits

9.1 **Why change?**

The conventional coil ignition system has been with us for about sixty years and was unchallenged until the ubiquitous transistor burst on the scene. It was admitted that the conventional system was far from perfect but we had learned to live with its weaknesses. One weakness is that the voltage delivered to the plugs falls off at higher engine speeds (see Fig. 9.1). Magnetos were

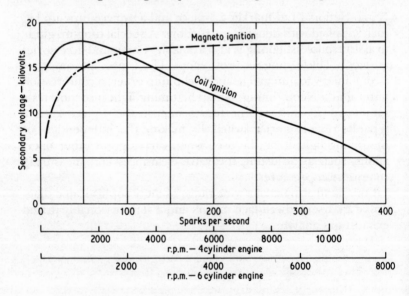

Fig. 9.1 Magneto ignition *v* coil ignition.

used in many early racing engines, since it was found impossible to design a coil ignition system that would give a high enough voltage to ensure reliable ignition on high compression engines at engine speeds above 6000 rpm.

The conventional ignition system has improved over the years and is adequate for most modern engines, *when in a good state of tune.* Unfortunately, it does not stay that way for long. After about 10 000 miles the plug gaps have increased and raised the demand level (see Fig. 9.2). Since the change in contact gap has also caused a falling off in coil performance the available voltage can become

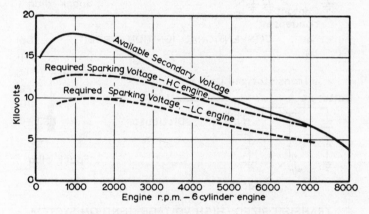

Fig. 9.2 Available voltage with coil ignition.

inadequate for the whole speed range. Misfiring at high engine speeds and poor starting are the usual symptoms. For this reason, it is usual to recommend that one re-gaps the plugs and adjusts the contact gap every 6000 miles. Since contactless electronic ignition systems do not have any contacts to wear they maintain good tune and require virtually no maintenance. The less expensive transistorized systems that retain the contacts suffer far lower rates of contact wear than the conventional ignition system since the spark erosion is much reduced, but they still suffer from heel wear.

9.2 **Transistor switching**

The simplest and cheapest electronic kits use a transistor or
thyristor to act as a relay switch to reduce the current carried by
the contacts (see circuit diagrams in Fig. 9.3). The current passing

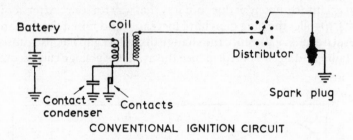

CONVENTIONAL IGNITION CIRCUIT

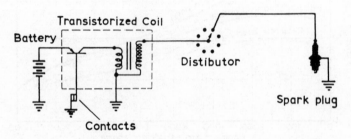

TRANSISTORIZED HIGH VOLTAGE IGNITION SYSTEM

Fig. 9.3 Circuit diagrams for conventional and transistor-
assisted coil ignition system.

across the contacts is reduced from about 4 A to as little as 0.25 A.
The function of the transistor is to step up the current passed to
the primary windings of the coil to a value of 6–7 A. By reducing
the current passed across the contacts arcing is much reduced and
their life is extended. Wear of the plastic heel of the moving
contact can still take place, and it is this factor that determines the
drift of the ignition timing and the change in dwell angle. Even so,
the life between tune-ups can be doubled.

A typical transistor-switching system is that made by Boyer-
Bransden Electronics Ltd.* This system at the time of writing

* Maker's addresses are given in the Appendix.

costs £13.11, including postage and VAT. At this price it will probably pay for itself in petrol savings alone in the first twelve months.

Transistorized ignition kits that retain the original contact mechanism are usually designed to magnify the *inductive* component of the spark discharge. A few systems, however, are designed to intensify the initial *capacity* component of the spark. Such systems are excellent for starting on cold mornings and will cope well with fouled plugs, having a relatively fast voltage rise time. Comparative tests of the two systems, however, suggest that those systems designed to magnify the inductive component are most effective in burning lean mixtures – and this after all is our most direct route to economical motoring. For this reason alone, we recommend for the medium-mileage motorist the Boyer-Bransden, the Mobelec Max or the Waso, three Inductive Discharge Systems priced below £25.

9.3 Contactless electronic ignition

Contactless systems are more expensive but are virtually maintenance-free. For high-mileage motorists they can be economically viable since the initial cost can be saved in reduced maintenance over a period of about two years.

9.3.1 Lucas OPUS System

Lucas have been supplying electronic ignition for Formula I racing cars for more than twelve years and their latest OPUS (Oscillating Pick-up System) is now available as an exchange distributor for a wide range of models. It is an expensive replacement, but for owners of older well-beloved quality cars, particularly those with high-revving 6, 8 or 12 cylinder engines the Lucas OPUS is probably the finest ignition system available today.

The latest Lucas system is contactless and has a spark capability of 800 sparks per second, twice that of the best contact-operated systems. It comprises four main components, an amplifier unit, a ballast resistor, an ignition coil and a distributor. The amplifier is a continuously operating fixed frequency (600 kHz) oscillator, transformer-coupled to an amplifier stage and an output stage which is a power transistor performing the function of the conventional contact-breaker. The amplifier unit is housed in a finned aluminium heat sink unit, since the power transistor must not be allowed to overheat. Those who are baffled by electronics jargon must accept that the amplifier is a 'super-contact breaker' that will never need any maintenance if it never overheats. Perhaps one in every half-million will fail. When it does the unfortunate motorist will be faced with a walk to the nearest phone-box.

The distributor (see Fig. 9.4) resembles the conventional distributor externally. Inside the cap one finds the customary rotor arm, but the familiar contact-breaker is replaced by the timing rotor which is a moulded fibre-glass-filled nylon disc with ferrite rods embedded in the periphery. The number of rods is equal to the number of cylinders. The pick-up module is in the form of an E-shaped transformer core, with the outer limbs of the E carrying input windings fed from the amplifier oscillator unit. The centre limb carries the output windings to drive the amplifier stage. By design, the resultant magnetic flux in the centre limb is negligible, hence the output signal is also negligible, when one of the ferrite rods is not in line with the pick-up module. When a ferrite rod passes the face of the E-core the magnetic circuits become unbalanced giving a relatively high signal voltage to the input coil. This signal, when fed to the amplifier, switches off the transistor. This breaks the primary circuit in the ignition coil, as in the conventional contact breaker. The voltage rise time is much faster than in the conventional distributor and a high voltage is produced in the secondary windings. Conventional vacuum and centrifugal advance mechanisms are incorporated in the distributor body.

9.3.2 The Mobelec Magnum Ignition

The contactless system made by Mobelec bears some resemblance to the Lucas OPUS system. A small plastic moulding is pressed over the distributor cam. As in the Lucas system, this moulding carries ferrite rods equal in number to the number of cylinders.

Fig. 9.4 The Lucas OPUS contactless electronic ignition system.

Labels on figure:
- Cover
- Vacuum timing control
- Electronic amplifier module
- Timing rotor with ferrite rods
- Pick up module
- Rotor arm
- Flashover shield

As these rods pass across the face of an inductive sensor a signal is fed to the electronic amplifier (See Fig. 9.5).

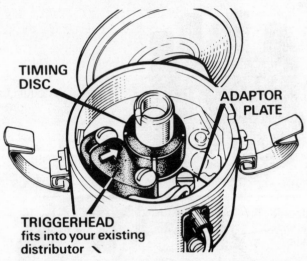

Fig. 9.5 Mobelec Magnum contactless electronic ignition system.

9.3.3 **The Piranha Micro-Logic Ignition**

In this contactless system, the coil discharge is triggered by a rotating chopper that 'makes and breaks' an infra-red light beam. The chopper is fitted in place of the standard rotor arm. Slots in the chopper uncover the light beam at the correct ignition timing. A light sensor then triggers the transistor-switched primary circuit of the coil.

9.4 **Reliability of electronic ignition**

Some early examples of contactless electronic ignition were not entirely reliable. Since the original contact-breaker had been removed a failure of an electronic component usually ended with the car on the end of a tow-bar. Some of the kits that retain the original contacts are provided with a switch to permit a quick change back to the conventional system. Even when such a switch is not provided it is a simple matter to change the leads back to the original positions. In fairness to the makers of the contactless systems the author must admit it is many years since he heard of a failure of such a system.

9.5 Economics

When tested by *Motor* (24 March, 1979) the relatively inexpensive Boyer-Bransden system showed an improvement in mpg of 2% at a steady 30 mph and 3.5% at a steady 60 mph. Overall, however, some systems showed gains, others showed losses, and the mean value for the seven systems tested was almost identical to those given by the standard Princess, the common test vehicle. It is interesting, even so, that the power curves measured on the dynamometer were higher *in every case*. This suggests to the author that the engine, when fitted with electronic ignition, could have been tuned with a leaner mixture than the standard engine and still give a similar power. At 4000 rpm the conventional O-series engine in the Austin Princess produced 78 bhp. With contact-assisted electronic ignition the power produced by the seven systems under test was increased by 3 bhp on average. The greatest increase was given by the Mobelec Max with an output of 83 bhp at 4000 rpm.

There was no doubt, from the evidence of these closely con-trolled *Motor* tests, that the more expensive contactless electronic systems gave a significant increase in mpg and a worthwhile increase in maximum power. The Mobelec Magnum system gave an improvement in mpg of 3.7% at a steady 30 mph and 4.5% at a steady 60 mph. The power at 4000 rpm was increased by 3 bhp. The mean mpg increase for the four contactless systems under test was 1% at 30 mph and 4% at 60 mph. The mean power increase at 4000 rpm was 4 bhp.

Based on the *Motor* figures, we could therefore anticipate a very small improvement in mpg from fitting a simple contact-assisted electronic ignition. At a very modest price of £13, however, the Boyer-Bransden system did actually show a worthwhile improve-ment. The author's measurements on his Austin Allegro 1100 confirm this. One positive gain with all such systems is a reduction in contact erosion.

The Lucas, Mobelec Magnum and Piranha systems are more expensive, but when fitted they reduce the overall maintenance costs for many years. With such systems one can predict an overall gain in mpg of 2–3%.

10

Bolt-on Aids to Economy

10.1 Gadgetry

The late W.C. Fields must have had the world of Advertising in mind when he advised, 'Never give a sucker an even break'. Every month we read of new gadgets that have a money-back guarantee and will give us more power, burn less fuel – even clean the carbon from the cylinder head! The French Government scientific officers were once hoodwinked into witnessing a demonstration of a car that ran on nothing but water. The Magic Circle probably know how the trick was performed.

The reader will understand why the list of economy aids in this chapter is so small. He must also accept that the author has no financial interest in any of them.

10.2 The electric fan

When gasoline flowed like water in the U.S.A., 7-litre sedans wasted as much power driving the fan at 70 mph as one of the smaller European saloons uses to propel the entire vehicle, i.e. about 25 bhp. The current models of cars all over the world are no longer designed to operate the fan continuously. Some are fitted with viscous clutch drives that only engage when a temperature sensor indicates that assisted cooling is required. At normal cruising speeds on the level, the air velocity passing through the radiator provides adequate cooling, requiring no assistance from the fan. Other designs use an electric motor to drive the fan. This motor only operates to drive the fan when a temperature sensor in the top radiator connection reaches a chosen temperature. This ensures a rapid engine warm-up and the fan is normally out of action. When the car is driven slowly in traffic in hot weather, driven hard on a motorway, or when climbing a steep hill, as the coolant temperature reaches the control temperature the fan motor is energized to supplement the cooling. Since, on a typical car, the fan is idle for 90% of the time the provision of this type of fan makes a useful contribution to economy.

10.2.1 The Kenlowe fan

The Kenlowe 'Thermatic' Fan has been in general use for at least twenty years. The magazine *Top Gear* reported an increase of 5 bhp in the peak power of a 1500 cc Hillman Avenger when tested on a rolling road dynamometer and a 1½ second improvement in the 0–60 mph acceleration time. *Practical Motorist* tested the Kenlowe Fan in 1979 on a Vauxhall Cavalier. There was a slight improvement in acceleration times, with improvements in mpg of 7.8% at a steady 30 mph and 9.5% at a steady 60 mph.

Current prices of Kenlowe Fans start at £28 plus carriage and VAT. Any motorist who covers more than 10 000 miles in a year will probably cover the cost of a Kenlowe Fan in his first year by the savings in his fuel bill.

It is not difficult for the average handyman to fit (See Fig. 10.1). No holes need be drilled. The temperature sensor can be fitted in the top radiator hose as shown in Fig. 10.2. The temperature setting of the control unit can be varied to suit the particular operating temperature of the radiator thermostat. The setting can be changed to suit winter or summer conditions.

10.3 Other gadgets

Devices to improve the performance or increase economy can be divided into four main groups: carburation; ignition; exhaust; and lubrication.

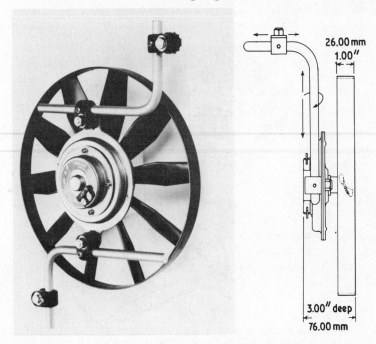

26.00 mm
1.00"

3.00" deep
76.00 mm

Fig. 10.1   The Kenlowe electric fan.

10.3.1 **Carburation**

About a quarter of a million GM Modifiers have been sold since
the late George Mangoletsi patented his manifold modifier about
thirty years ago. Mangoletsi had a wide experience of the behaviour of wet fuel/air mixtures in all configurations of manifold
and the elegant device he eventually marketed has been shown to
be as effective as the best, usually very complicated, re-entrain-
ment gadgets patented by other engineers.

The carburation weakness he helped to alleviate is one that is
present in varying degrees on all engines fitted with carburetters.
With petrol injection, normally injected at the valve port, the
problem does not exist. With low air velocities through a carburetter venturi, and particularly when the engine is cold, the atomization is poor. Under these conditions many large droplets of
petrol are created and these tend to be thrown on the walls of the

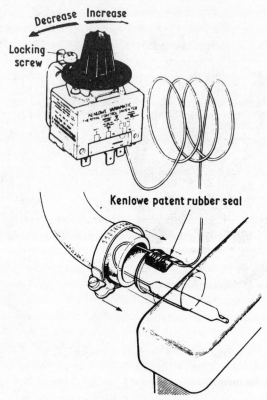

Fig. 10.2 The Kenlowe temperature sensor.

manifold. On unheated or poorly-heated manifolds the walls are
covered with a liquid film that is slowly moving towards the inlet
valves. As this liquid film reaches the branches in the manifold,
there is a tendency for the major portion of this moving film to
pass into the more centrally placed branches, resulting in a leaner
mixture reaching the outer cylinders.

The GM Modifier, shown in Fig. 10.3, is a simple inlet restrictor
fitted in the manner of an additional gasket between the carburetter flange and the induction manifold. The restrictor only
reduces the cross-sectional area by about 10% and is designed to
form a collecting channel for the liquid fuel with a sharp-edged

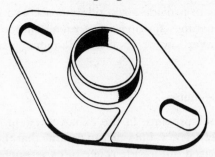

Fig. 10.3 The GM Manifold Modifier.

inner lip from which the fuel can be re-entrained into the air-stream. An additional feature is a small tangential air bleed to lean off the mixture strength. This will have a negligible effect at large throttle openings. GM Modifiers are available for most makes of carburetter, including the 4-stud flange Webers.

Many tests reported by the motoring press show a slight improvement in overall mpg from this device. The latest *Motor* test (29 October, 1979) showed no improvement. It has been the author's experience that the benefits of the GM Modifier are usually apparent on a frosty morning when a manual choke can be pushed fully closed much sooner when this device has been fitted. Even if the device gives no improvement in steady speed fuel consumption tests *with a warm engine,* a reduced use of the choke must give a small overall increase in mpg.

10.3.2 **Ignition**

The advantages of electronic ignition systems have been well covered in the previous chapter. Many, many spark intensifiers have been advertised in the motoring press, even in the daily press, during the author's lifetime. One audacious company actually marketed such a device using wood screws screwed into wood bungs fitted into the ends of a short length of plastic tube. The H.T. spark jumped the gap between the ends of the wood screws, thus providing a secondary spark gap. Even before the author saw the light of day motorists had used mother-of-pearl shirt buttons to provide a secondary spark gap for recalcitrant engines, sometimes starting a fire under the bonnet when the carburetter overflowed. There is little merit in such devices.

10.3.3 The exhaust

Excessive back-pressure in the exhaust system can result in a decrease in efficiency, but one can assume that the car manufacturer has carried out a wide range of experiments before finalizing his exhaust layout. In general, then, if we are not to fall foul of the Law by producing too much noise we must retain the standard exhaust system. In many systems, the removal of the second silencer (often called a resonator) will result in a very sporting boom at certain engine speeds. The overall back-pressure would be reduced by its removal, but the improvement in economy would be marginal.

10.3.4 Exhaust extractors

Many firms have offered us devices to fit on the end of the tail pipe that have been designed to use the forward motion of the car to create an extractor or venturi action on the gases leaving the tail pipe. The Cornette extractor has been sold by the million and is the most popular of all such devices. As shown in Fig. 10.4 a forward-facing funnel presents a wide air entry to a divergent passage which gives an increased air velocity in the annular passage around the end of the tail pipe. This increase in velocity results in a small extractor action on the end of the tail pipe. The latest tests by *Motor* (20 October, 1979), carried out with their usual professional accuracy, showed a gain in overall consumption of 0.7% when a Cornette exhaust ejector was fitted to a Datsun 140Y Coupe.

Extravagent claims have been made for many of these exhaust extractors and have even been supported by the evidence of tests carried out by some of the more slap-happy 'fringe' motoring magazines. In the author's opinion, the only superiority shown by these more elaborate devices over the Cornette is in their price. They are usually very expensive.

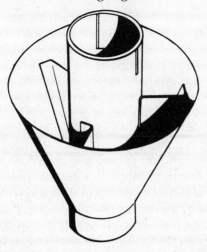

Fig. 10.4 The Cornette exhaust ejector.

10.3.5 Lubrication

Only when an engine has reached that terminal condition best described as an 'oil-burner' does it ever pay to use the very cheapest straight-grade oil. Even then it is foolish to prolong the agony. An oil-burner is not only a public nuisance; it is an expensive form of transport. A typical engine of 1500 cc capacity in good condition will use oil at a rate not exceeding a pint every 500 miles. For the first 20 000 the oil consumption should be even less and it would be very foolish to use anything but the premium grade of oil as sold by such long-established suppliers as Castrol and Duckhams.

Viscosity Index

The straight-run motor oils we used thirty years ago were no less effective as lubricants than the latest products from the big oil companies. They had a greater tendency to form sludge-like emulsions in the sump and carbon deposits inside the combustion chamber and required frequent oil changes to maintain the engine in good condition. Anti-oxidant additives in modern oils have helped to extend the life between oil changes. High viscosity at low temperatures was always a problem with the old straight-grade oils. A relatively light oil, i.e. SAE Specification Grade 20, would, in many engines become far too 'thin' at the working temperature of a crankshaft bearing under full load. Oil pressure would fall too low and too much oil would escape at the end faces of the bearings. On the majority of engines in the Vintage era the recommended oil grade was SAE 40 or 50. These engines had starting handles, but the oil-drag on a cold morning was enough to bring on a heart attack in the middle-aged. Viscosity Index was established by American lubrication technologists as a means of comparing the ability of some oils to show a lower rate of change of viscosity for a given change of temperature. Pennsylvanian oil was known to rate high with a relatively low change in viscosity. Asphaltic oils from the Gulf region were known to be poor. An arbitrary scale based on these oils was established and is now used world-world as 'the Viscosity Index'. There still remained a crying need for oils with an acceptable viscosity at the extremely low temperatures that are encountered in places such as Alaska and Siberia.

The breakthrough came when the lubrication chemists developed synthetic additives called 'viscosity index improvers'. These are the additives that gave us the modern multi-grade oils. These oils are specified in terms of their equivalent SAE grades at different points on a temperature scale. For example, a 10W/40 Grade oil has the viscosity of the former SAE 10W (W for Winter) oil at 0°F and 40 Grade oil at 300°F. The advantages of using a multi-grade oil are obvious. During engine warm-up, the gain from the reduced viscosity, compared with a straight 40 Grade oil, will amount to an increase in overall mpg of 2–3% in winter conditions.

Proprietary additives

'Super-lubricants', such as STP, Redex and Molyslip will be well-known to the reader, since these products are so well advertised. A modern engine in good condition, lubricated with a premium multi-grade oil, will gain little of value from the use of such additives. Some oil companies have even suggested they can upset the delicate balance of the additives already present in their multi-grade oils. This could be true, but the oil companies have sometimes made rather extravagant claims for their own products. In all fairness, the author must mention one possible condition when the use of a 'super-lubricant' could be of benefit. Sometimes an engine designer makes a mistake, sometimes a production department fails to choose the right material or the right surface treatment for a sensitive component. One popular British car had a tendency to suffer excessive wear of the cam lobes. The addition of STP, Redex or Molyslip or some similar concoction, would probably extend the life of the camshaft in such a case, since these products contain additives to reduce the scuffing action when normal lubrication breaks down.

We must conclude with a confession. The author has used Molyslip and Redex on several cars during the critical running-in period on new cars. He cannot prove they helped, but at least they did no harm.

11

Driving for Economy

11.1 **Filling the tank**

Obviously, we must take advantage of any price wars that develop
from time to time between the big petrol companies and occasionally between neighbourhood filling stations. We are particularly
fortunate if our cheapest filling station is about a mile from our
home on our way to work in the morning, since we can achieve
two economies at the same time. The first economy is derived
from the 'heat-soak' effect on the induction manifold. During the
colder six or seven months of a typical British year one will often
travel one and a half to two miles before the manual choke can be
pushed completely OFF. However, as the car is standing at a
filling station near home with the engine switched off, heat will
be soaking from the hot exhaust manifold into the induction
manifold. Quite often after this heat-soak the car can be driven
off without the use of the manual choke. An automatic choke
often reacts in the same way, thus making a small economy.

The second economy is a little more substantial and never
seems to be mentioned in the Motoring Press. The diurnal temperature variations of the petrol in the underground tanks at a
filling station are not as large as the diurnal atmospheric variations, but they do exist and we can benefit from them. Modern
filling stations have 5000 gallon tanks and the temperature at 8
a.m. will be about 3°C lower than at 6 p.m. The density of the
petrol will therefore be about 1% greater in the morning. The
petrol pumps meter on a volumetric basis, i.e. in gallons or litres.
We therefore receive a slightly greater *mass* of petrol at an early

morning filling. Since the available energy in a fuel is based on a known value *per unit of mass* the gain is indisputable. If, of course, we are late for a morning appointment and there is enough petrol left in the tank we would postpone our re-fill until the evening. This petrol would cost us about 1.3 pence per gallon extra in real energy terms. Taken on an annual basis, however, regular evening fill-ups would cost us about £8 per annum more since a typical commuter spends about £650 per annum.

11.2 **Petrol grade**

The R.O.N. (Research Octane Number) of any star grading is permitted to vary by plus or minus 2 numbers and a sensitive ear can detect these differences when a particular engine is operating close to the knock limits of the particular grade. The sharp metallic pinging sound we usually call 'pinking' will usually occur when accelerating hard up a gradient, especially with a hot engine. An occasional pinking is not a serious matter. If, however, the pinking develops into a more persistent knocking sound approaching the noise levels of the familiar Diesel knock we can anticipate trouble; possibly piston or exhaust valve failure.

The message then is two-fold:

(1) Operate on the car manufacturer's recommended star grade. The use of a lower grade is likely to be false economy.
(2) Listen for the occasional pink when accelerating hard or climbing a hill on full throttle. If any particular brand of petrol tends to produce pinking, while others do not, the answer is simple. It is not necessarily a condemnation of the particular brand. It could simply mean that this brand does not perform well in your design of combustion chamber.

11.3 **Driving techniques**

Driving for economy can be cultivated into a fascinating hobby. In recent years it has almost become a sport. Paradoxically, hundreds of gallons of petrol are wasted, or so it would seem, while dozens of professional and amateur drivers traverse the country going nowhere in particular while competing in Economy Rallies. My old colleague, Joseph Lowrey, now a free-lance motoring journalist, will no doubt bristle with indignation if he ever reads this. It is logical, he would say, to regard the Economy Rally as a facet of road research leading to valuable savings in our national fuel bill. He could be right. Anyone who can tune a car and then drive it to give nearly double the mpg figures given by ordinary motorists must be a serious research worker.

Economy driving is largely common sense. Few of us need to be told that Grand Prix starts from traffic lights waste fuel. High speed driving is also wasteful. Government fuel consumption tests are carried out at two constant speeds, 56 mph and 75 mph. The gain in mpg by travelling at the more modest cruising speed is more pronounced with the lower powered cars. A Ford Fiesta 1.1 at a constant 75 mph gives a reading of 33.6 mpg, increasing to 47.1 mpg at 56 mph. A Ferrari 308 GT4 will give 19.2 mpg at 75 mph, only increasing to 22.3 at a steady 56 mph. The smaller engined car gains in two ways by running at the lower speed. The air resistance is only about 55% of that at 75 mph. Moreover, at 56 mph, the engine is running very close to the peak of the torque curve (47 mph in top gear). Even though the losses from air drag are almost halved at the lower speed the Ferrari is operating on a small throttle opening at a constant 56 mph. The efficiency of the engine is therefore much higher at the higher engine speed in the case of a high-geared sports car. But why should we worry about the rich! Very few owners of Ferraris and Lamborghinis will be reading this book.

Every car has a *reasonable* economic cruising speed. With the smaller-engined European cars, absolute economy is achieved at a very low speed in top gear. Some exceed 60 mpg at a steady 30 mph, but this is not acceptable as a cruising speed on a long journey. A reasonable economic cruising speed is usually found to be very close to the engine speed at which the torque curve peaks. The point on the graph of maximum torque against engine rpm where the torque curve reaches a peak value is influenced by such parameters as the valve timing, the size of the valves, the combustion chamber design and the carburetter venturi size. All these values have been chosen by the designer to suit the particular car. For economic motoring we should aim to keep the engine speed within a band that embraces 20% below maximum torque rpm and 20% above. As an example, the author's Allegro 1100, with a torque curve that peaks at 2900 rpm should be driven to maintain the engine speed between 2300 rpm and 3500 rpm. In top gear, this range corresponds to speeds of 35 and 55 mph. In third gear the speeds range from 25 mph to 38 mph. By good design, third gear acceleration takes the car to the peak of the torque curve (31 mph) leading to 30 mph in top gear, the usual legal limit in town traffic.

11.4 **Hills**

The driver can use his knowledge of the torque peak on his particular car to choose the correct gear when climbing a hill. If, for example, the Allegro 1100 falls in speed on a hill in top gear to about 35 mph and the top of the hill is not in sight a downchange will probably drop the road speed to about 33 mph, but the engine torque and efficiency will now be increased in this lower gear. With the increase in torque the 35 mph will soon be re-gained and, if the gradient does not increase, this speed will be maintained easily with a slight reduction in throttle opening. If, however, the driver had decided to slog along in top gear the engine would be operating on a falling torque curve and the speed would fall well below 33 mph. This is not only uneconomical, but is not the kindest treatment for the crankshaft bearings.

To determine the most economical hill-climbing speeds in the various gears a driver needs to know the following three things about his own vehicle:

(1) The engine speed for maximum torque.
(2) The road speed in top gear at a given engine speed. This is usually expressed as mph per 1000 rpm.
(3) The gear ratios.

The first item, if not given in the driver's handbook, can usually be found in *Motor* or *Autocar* Show Numbers or in copies of their Road Test Reports on the car in question. Items 2 and 3, if not found in the driver's handbook will be found in the maker's Service Manual or some well-documented publication such as the

Workshop Manuals published by Haynes.

The calculation is very simple. The road speed giving maximum engine torque in top gear is found by multiplying item 1 by item 2. Item 3 can then be used to convert this value to the appropriate one for the lower gears by dividing by the gear ratio relative to the top gear ratio.

To save petrol one should avoid accelerating up a hill. With experience on a known route one learns when to increase speed slightly at the approach to a hill. This should give sufficient momentum to top the rise in the appropriate gear *within the known economic speed range for the gear*. On the Allegro 1100 for example if we know that a drop to third gear will be necessary we would try to climb the hill within the speed range 25–38 mph.

In an undulating terrain it pays to gain speed downhill and lose speed uphill. Naturally, the experienced driver avoids runaway speeds on steep hills.

11.5 **Anticipation**

In less crowded times and before we had a speed limit the author sometimes drove at high speeds, occasionally at speeds above 'the ton' for mile after mile. More than anything this reveals the author's age! The secret of survival with no help from a well-planned Motorway system was anticipation. With good eyesight we could spot trouble at least a quarter-mile ahead and begin to apply the brakes. In the same way we can save petrol in these changed times by anticipating delays and *not applying the brakes*. Every application of the brakes means a loss of energy. You should try to keep a distance behind any tight pack of vehicles that has formed a log-jam behind a slow-moving vehicle. The first action on spotting the hold-up ahead is to ease the throttle back sufficiently to coast up to the procession at the actual speed of the procession, thus avoiding any use of the brakes. Inevitably, you will meet overtaking clots who will pass you and rob you of valuable slowing-down space. Notice how they drive right up to the tail end of the procession, hoping it will evaporate, then brake hard at the last moment. Road hogs often save as much as ten minutes on their daily journey to work, then waste fifteen minutes at the office drinking coffee to calm their nerves while boring their fellow workers with hair-raising stories of the idiots they have encountered en route.

Economical driving is achieved by saving the brakes for emergencies, by maintaining the normal cruising speed on the majority of bends, only reducing speed before a bend if it appears to be too acute or too slippery and finally by avoiding hard acceleration and high speeds.

One cannot usually switch off the ignition at traffic lights, but there are occasions when long delays are apparent, i.e. at light controlled road works, when a small economy can be made by switching off the engine.

11.6 Tyres and pressures

Nearly all cars today are fitted with radial tyres. The older crossply are to be avoided. They not only wear out more quickly but they have a slightly higher rolling resistance. The recommended maker's tyre pressure should be maintained, making weekly checks with a reliable tyre gauge. One can use higher tyre pressures in an economy rally to reduce rolling resistance, but the use of high pressures all the time reduces the contact patch area and the safe cornering speed. An uneven wear pattern across the width of the tyre tread can also be expected from continuous overinflation.

11.7 Maintenance

Engine tuning has been well covered, but the neglect of the chassis can also waste fuel. Binding brakes or worn wheel bearings can be checked by jacking up each wheel in turn. Incorrect alignment of the front wheels is sometimes indicated by feathered edges to the tread pattern. Poor front wheel tracking not only wears out tyres rapidly; it also wastes petrol.

11.8 Envoi: so what have we saved?

Good tuning
The tests quoted in Chapter 1 suggest that the majority of cars in

Great Britain are wasting about 14% of their petrol since their owners neglect to tune them. With a careful re-tuning every 8000 miles the falling off in mpg at the end of the period should not exceed 6% – an average reduction of 3%. We can therefore forecast an average improvement of 11% from regular tuning.

Tank filling

A typical motorist who changes from a random time for re-filling his petrol tank to a regular morning re-fill can gain about 0.5% in improved mpg.

Electronic ignition

(a) Since a contact-assisted electronic ignition system should suffer less contact wear over a period of 8000 miles we can assume a reduced average loss of tune, about 2% instead of 3%. This represents an effective gain of 1%.
(b) A contactless electronic ignition system will give the same effective gain of 1% from the reduction in the gradual loss of tune between tune-ups. In addition to this, there will be an overall improvement in mpg of about $2\frac{1}{2}$%, i.e. a total gain of $3\frac{1}{2}$%.

Thermostatic fan

There is no doubt that this is the most cost-effective device we can fit to a car. That the majority of car manufacturers are prepared to fit one as original equipment is proof enough that this equipment is a good investment. An overall improvement of about 5% in mpg will be given on a typical medium-powered car.

Driving technique

An average driver who seriously adopts our recommended economy technique should get an improvement in mpg of at least 5%. This modest improvement is based on an acknowledgment that the reader of this book is already driving with economy in mind. Deep down in his sub-conscious he probably yearns to be a lead-footed rally driver, but the state of his bank balance keeps his foot off the accelerator.

If the reader cannot find the ready money to buy electronic ignition or a Kenlowe fan he can, for a modest outlay of about £20 for tuning equipment, keep his engine in a good state of tune.

The predicted savings in this case would be 11% + $\frac{1}{2}$% + 5%, i.e. 16–17%. Even on a modest annual mileage of 8000 miles this would still amount to an annual saving of about £70.

Total savings

		Totals
Regular tuning	11%	11%
Morning tank-filling	$\frac{1}{2}$%	$11\frac{1}{2}$%
Contactless electronic ignition	$3\frac{1}{2}$%	15%
Thermostatic fan	5%	20%
Driving technique	5%	25%

Appendix

Suppliers' Names and Addresses

Carburation modifier

GM Modifier   GM Manifold Development Ltd., Lymm, Ches.

Carburetter service kits

Stylex (all types of carburetter), Hollands Road, Haverhill,
  Suffolk, CB9 8PU.
SU-Butec, Dormer Road, Thame, Oxon.

Carburetters, spares and information

SU Carburetters, Wood Lane, Erdington, Birmingham,
  B24 9QS.
Solex, Zenith, Stromberg   Zenith Carburetter Co. Ltd., Honeypot
  Lane, Stanmore, Middlesex, HA7 1EG.
Weber   Phegre Engineering, Hartley Wintney, nr Basingstoke,
  Hants; Radbourne Motors (London) Ltd., 8 Bramber Road,
  London, W14.; Merchiston Motors, Merchiston Mews, Edin-
  burgh, 10.

Exhaust extractor

Exhaust Ejector Co Ltd., Wade House Road, Shelf, nr. Halifax,
  HX3 7PE.

Fans, electric

Kenlowe Accessories and Co Ltd., Burchetts Green, Maidenhead, Berks., SL6 6QU.

Smith Industries, 50 Oxgate Lane, Cricklewood, London, NW2 7JB.

Handbooks

Autobooks Ltd., Golden Lane, Brighton, BN1 2QT.

Haynes Publishing Group, Sparkford, Yeovil, Somerset, BA22 7JJ.

Murray Book Distributors, 107 Fleet St, London, EC4.

Ignition, electronic

Bosch   Bosch Ltd., Rhodes Way, Watford, Herts., WD2 4LB.

Boyer-Bransden   Boyer-Bransden Electronics Ltd., 46a Apsley Rd, South Norwood, London, SE25 4XT.

Lucas   Lucas (Sales and Service) Ltd., Gt Hampton St., Birmingham, B18 6AU.

Mobelec   Mobelec Ltd., Oxted Mill, Spring Lane, Oxted, Surrey.

Piranha   Piranha Ignition, Kenworthy House, Freckleton Street, Blackburn, Lancs., BB2 2EL.

Sparkrite   Electronic Design Associates, 82 Bath St., Walsall, WS1 3DE.

Spitfire   Creelfine Automotive Development Ltd., Manufacturing Unit 1, Priory Rd., Aston, Birmingham.

Surefire   Suretron Systems (UK) Ltd., Piccadilly Place, London Road, Bath, BA1 6PW.

Waso   Waso Ltd., Whiteway Rd., Queensborough, Kent, ME11 5EQ.

Tuning aids

Colorplugs   Gunson's Colorplugs Ltd., 66 Royal Mint St., London, E1 8LG.

Jetsetter Patent Enterprises Ltd., 143–145 Kew Rd., Richmond, Surrey, TW9 2PN.

Manifold adaptors & T-pieces Speedograph Ltd., Darlton Drive, Arnold, Notts., NG5 7JR.

Tuning instruments

Command Narco National Ltd., St Mary's Works, Krooner Rd., Camberley Surrey, GU15 2QY.

Crypton Crypton Ltd., Bridgwater, Somerset.

Gunson's Gunson's Colorplugs Ltd., 66 Royal Mint St., London, E1 8LG.

Suntester Suntester Ltd., Oldmeadow Rd., King's Lynn, Norfolk, PE30 4JW.

Index